Performance Corvairs

How to Hotrod the Corvair Engine and Chassis

Seth Emerson and Bill Fisher

California Bill's
Automotive Handbooks
Tucson, Arizona

Publishers
Howard Fisher
Helen Fisher

Editor
Howard Fisher

Cover Design
Gary Smith, Performance Design

Text Design and Production
Doug Goewey

Published by
California Bill's Automotive Handbooks
PO Box 91858
Tucson, AZ 85752-1858
(520) 547-2462
www.californiabills.com

ISBN-10 1-931128-22-7
ISBN-13 978-1-931128-22-3
Ebook 978-1-931128-37-7

Printed in the United States of America

3 4 5 6 7 23

Front Cover
1966 Corsa silver coupe, photo by Ned Madsen, Bush Stadium, Indianapolis, Indiana, 2008.

1965 Corsa engine compartment photo by Seth Emerson.

Back Cover

John Egerton's #6 white 1964 Monza, photo by Rick Norris. Detroit, Michigan, 2007.

Seth Emerson's #46 yellow and black 1965 Yenko Stinger, photo by Dito Milian, gotbluemilk.com. Motorsports Photography. Thunderhill Raceway, Willows, California, 2012.

Dave Edsinger's #18 white and blue 1966 Yenko Stinger, photo by Rick Norris. Detroit, Michigan, 2007.

Seth Emerson with Corvair-powered Lola T320, photo by Silicon Valley Corsa, Oakland, California, 1993.

Road America 2016, photo by James Rice.

Title Page
Seth Emerson #46 1965 Yenko Stinger, photo by Dito Milian, gotbluemilk.com. Motorsports Photography. Thunderhill Raceway, Willows, California, 2012.

Corvair Society of America
CORSA is a world-wide organization with 4,800 members and 125 local chapters worldwide. Founded in 1969, it is a large and enthusiastic group of automobile enthusiasts and one of the best single marque clubs in the world. The organization's monthly magazine is called the *CORSA Communique,* www.corvair.org

Notice: The information in this book is true and complete to the best of our knowledge. It is offered without guarantees on the part of the author or California Bill's Automotive Handbooks. The author and publisher disclaim all liability in connection with the use of this book.

Table of Contents

About the Author

Bill Fisher

Author Bill Fisher got involved with cars early on. His grandfather owned Chevrolet agencies in Illinois and from age six he played behind the garage stringing old auto batteries together to produce more power.

His publishing career began in 1947 while pursuing a master's degree in education at the University of Southern California. As it came time to write his Master's thesis, he decided to combine his love of automobiles and English. He wrote *How to Construct a Hot Rod for High School Auto Shop Classes*. This thesis was the basis of his first published automobile book. Later, in 1948, he had a speed equipment shop in Eagle Rock, California and supplied speed parts to racers competing at Bonneville, El Mirage, and Muroc Dry Lakes. This was in addition to his flourishing mail-order automotive book business. Parts as well as books were sold to hot rodding enthusiasts across the country.

Bill was a huge fan of Chevrolet inline six-cylinder engines and when the Corvair came out in an opposed-six configuration he dug in to learn all he could.

In 1963, he founded HP Books (HP for horse power) and authored the original *How to Hotrod Corvair Engines* in 1964, revising it twice by 1969.

He loved motorcycles and cars. He rode them, tuned them, hot rodded, and raced them at the dry lakes. In 1981 he became a member of the Bonneville 200 Mile Club, breaking a record in his C Coupe at the Bonneville Salt Flats.

Acknowledgments

Bill Fisher passed away in May, 1999. These are his acknowledgments from the 1969 Revision III of *How to Hotrod Corvair Engines.*

No book of this type is ever produced single-handedly. I am deeply indebted to the many men and organizations who have been lavish in their assistance, both in the writing of the original edition, and now in this revised version. The Technical Projects Department of Chevrolet Public Relations has provided outstanding help in answering technical questions, and in providing technical information and photographs that have been used throughout this book. Chevrolet's Merchandising Department has also been very helpful. GM policies prevent naming these gentlemen, but I am most grateful to each of them.

Over two years were spent in creating the original edition. This revised edition represents continuous digging and research into the use of Corvairs since the original was published. Encouragement from Frank McGurk of McGurk Engineering and Jerry Light of Vic Hubbard Automotive were especially helpful in keeping the original project underway. Three men have been especially helpful in providing information on their competition experiences with Corvairs: Don Eichstadt, Doug Roe, and Ted Trevor. They have helped me to make this revised version better in dozens of details. Special thanks is due to several friends: Angus McDonald for the original cover design, John Joss for editing advice, and C.E. Camp for invaluable technical advice on various aspects of mechanical engineering relating to high-performance engines.

About the Author

Seth Emerson

I bought my first Corvair in 1967, after exploding a clutch in my latest/last 1955 Chevy. My parents okayed a "Chevrolet OK Used Car" purchase, and even co-signed for the loan, as long as it wasn't a '55 Chevy. They figured it was so different than the other cars I had been modifying that I would leave it alone and concentrate on college instead. Sorry, Mom! I think I stopped in and bought an IECO catalog on the way home! In the last 45 years, I have raced Corvairs in every type and class of racing in which they run (except off-road racing), including SCCA autocross and road racing, drag racing, and rallying. I have driven Corvairs as street cars for many years, though not as my only car. My Corsa convertible still has the engine that my girlfriend helped me build. I am so thankful that I married her, 40 years ago. In the late '70s, I responded to friends' laments about the poor quality ignition wires available for their race cars by starting Silicone Wire Systems. I have supplied thousands of 8mm silicone ignition wire sets to most of the Corvair Parts vendors. I joined The Corvair Society of America (CORSA) in 1973, just missing the Anaheim, California, convention. By 1980, I was fully engaged and I haven't missed a CORSA convention since. I was a Corsa director for more than fifteen years and President of Corsa twice. I have driven

Corvairs to FTD trophies at conventions and placed a car into Senior Division Concours status as well. In short I have lived the Corvair life deeply over the past 45 years. I don't plan on stopping soon. And I will continue to research new ways of applying technology to our older cars. Why don't you join me?

Acknowledgments

As Bill Fisher said many years ago, a book like this cannot be produced by one person. First Bill and then Howard Fisher supported my efforts in the update. Bob Helt and Richard Finch, both prolific Corvair authors, also encouraged me to complete the work. A big thank you goes out to those who are still building Corvair race cars and putting them on the track in front of an often skeptical public. The Corvair hasn't quite outlasted all of its critics, but it is well on its way to doing just that. Every time a Corvair puts the hurt on another car at the track, at the autocross, or even on the street(!), it validates Ed Cole's concept, Bob Benzinger and Maurice Olley's design, and countless Corvair developers from Bill Thomas to Doug Roe to Don Yenko to John Fitch.

To my Corvair friends, Bob Coffin, Chuck Sadek, Dave Edsinger, Rick Norris, Jim Schardt, Eddie Meadows, Warren LeVeque and a dozen others, your perseverance has always been my inspiration. Not only did you share your hot rodding techniques, you shared your cars with me at the track. I cannot thank you all enough, this is about the best I can do. My wife, of course, put up with my endless Corsa events, track trips, hours in the garage, and various amounts of money stuffed into Corvairs and "parts." Thanks Barbara, and thanks to you all. I hope you enjoy reading the book as much as I enjoyed writing it.

Introduction

The original *How to Hotrod Corvair Engines* (HTHCE) book was released in 1964. "California Bill" Fisher, a long time hot rodder and author had been enamored with the Corvair for some time. (Undoubtedly through his friendship with Ted Trevor, owner of Crown Manufacturing.) Bill had completed an engine swap of a Corvair motor into his Porsche 356 and was about to buy and drive a 1966 Corvair as his daily driver. The company that released HTHCE was called HP Books. The Corvair book was HP Book number one. The success of that book spawned a publishing empire that went on to publish hundreds of different high performance automotive titles, created a crockery cookbook series, and finally every other type of book for do-it-yourselfers. Ownership of the company, HP Books, changed hands a few times, but Bill, along with his wife, Helen; and son, Howard; continued to release books on varying subjects, including cars and car performance. The last revised issue of HTHCE, done in about 1973, was reprinted a few times by Corvair vendors, and hasn't been available for many years, except at swap meets. Back in about 1993, Bill and I agreed that a new revision of the book was needed. Because I was eager to write and he was busy with other projects, we agreed that I would update and rewrite the book, gathering in 20 more years of experience. The 20 years of experience has turned into 35, as other activities, both Corvair and otherwise, displaced much of my writing time. Progress was being made, however. And at the time of Bill's death in 1999, Howard Fisher, his son, and I agreed that I should continue the effort.

Ironically, the Corvair, for a car 50 plus years since first production, is still inspiring authors to produce books about it. As the cars age, authenticity becomes harder to achieve and preserve. When it debuted, the original HTHCE was the only book about Corvairs and the book covered the year-to-year differences in detail, at least from the performance point of view. Since that time, many books have been released to document what Chevrolet produced and how they went about building the cars. In this latest book, much of that authenticity data has been minimized, unless needed to indicate parts differences or sources. In short welcome to *Performance Corvairs—How to Hotrod the Corvair Engine and Chassis*.

1

An Unusual Automobile–Yearly Changes

Although Detroit's major industry was always quite secretive about new-car features prior to introduction, it is doubtful if any automaker had come close to the cloak-and-dagger methods GM used prior to Corvair's introduction. Workers on the preliminary effort removed all Chevrolet insignia from their clothes and tools so Chevrolet's name would not be connected with the project on which they were working. All material and tooling bids came through Holden, GM's Australian branch, so other makers would not connect Chevrolet with this new car. The first test cars used Holden nameplates as a further deception. As a result, Corvair really surprised America when it was introduced on October 2, 1959.

The planning and effort that Chevrolet put into the construction and manufacture of the Corvair indicates that a long production life was planned. The research and development efforts, which were ongoing throughout the production life of the Corvair (to a point), also lay credence to those plans. The fact that the Corvair was produced for about ten years would shock the automotive production planners of today, when major models are introduced to the market, their successors are already on today's computer design systems, and a new model that doesn't score fast is doomed to quick extinction.

Corvair was to fit into the compact-car market as a six-passenger car with acceleration, ride and handling equal to that of the larger six-cylinder Chevys. In some ways it surpassed that goal, but the cost targets that GM engineers were held to resulted in a pretty plain-Jane car. Suspension decisions, notably the non-inclusion of a front sway bar as standard equipment, caused dissension

This photo shows the important parts of the 1960 Corvair engine. This book tells you about the minor and major modifications you can make to this engine for greatly improved performance. Although the engine borrows heavily from European designs, it offers many innovations which the automotive enthusiast can appreciate.

within Chevy's engineering community. The compact car market was seen as an economy-car market and the Corvair was to squeeze 25 MPH from a gallon of gasoline at 50 MPH. GM was always looking at promoting their new models and displayed an up-market Corvair at a 1960 auto show in Chicago. Show attendees were enthusiastic about the car and GM noticed. So, the 1960-½ Monza line was introduced with extra-cost HP and a pretty bucket-seat interior. Buyers proved eager to lay down more money for additional frills and performance. Astutely, GM quickly threw out the economy, plain-Jane approach and changed Corvair's image to appeal to those regarding it as a "sports car." The detailed changes made each year's new model notably better in these respects than the 1960 models. Even so, the basic body was retained for five years with 1965 marking the first and only major styling change during its lifetime.

Corvair had a solid market acceptance. This accounted for 1.25-million sales by July 1964. There were several reasons for it. These included a whispering campaign among enthusiasts that this was "A poor-man's Porsche," and a general revolt against the sameness of other American cars. The Corvair was "different" with a host of fascinating technical features, including being the only rear-engined car ever to be mass-produced in the U.S. These marketing strengths were also marketing weaknesses. Many people had to be convinced that the Corvair offered enough benefits for them to embrace the radical rear-engined design.

As the British magazine *Autocar* declared in its January 1961 road test, "During the decade that has just passed, surely no new car at its introduction so stirred the whirlpool of rumor and counter-rumor, praise and ridicule, as did the Chevrolet Corvair when it arrived on the motoring scene 15 months ago. In four major respects it represents a complete departure from North American practice; its six-cylinder engine is horizontally-opposed, air-cooled, placed at the rear of the car, and all the wheels are sprung independently."

This marked difference between the Corvair and almost every other car of that era had earth-shattering effects on the American automobile market. Many people now feel that the explosion of the "sporty" car market, exemplified by the Mustang, Camaro, and countless others, owes its very existence to the introduction of the Corvair Monza models. Lee Iacocca, the "father" of the Mustang admitted the Monza was his inspiration. In another, perhaps less flattering effect, the uniqueness of the Corvair, the history of the media attention, and even the various lawsuits may have started the U.S. government's forty-year foray into the car safety field.

If you really crave a complete understanding of Corvair design, get a copy of SAE Paper 140C, "The Chevrolet Corvair." *Car Life* magazine, now departed, had a Corvair issue that contained vast quantities of design information from the SAE paper, plus important observations from its own technical staff. Reprints of this issue were introduced to the market several years ago. As a Corvair owner, you may be most interested in learning about some of the obstacles that were imposed and surmounted in placing the engine at the rear. The engineers knew that a true compact should not have any hump in the floor area as this reduces comfort for the middle passengers. Consequently, they investigated both front engine/drive and rear engine/drive cars before choosing the latter. Given the production constraints of the late 1950s, GM felt that the rear-engine car gave handling characteristics closer to those expected by the typical U.S. motorist. This demanded an engine which would be lightweight to permit good weight distribution when mounted behind the rear axle, and one which would fit into the usual "trunk space." Engine height was critical, hence the unusual flat-fan location. This design also resulted in the least handling change when the car was fully loaded. A Corvair promotional film "The Case of the Methodical Men" outlines the initial design, testing, and production of the Corvair.

Unusually high amounts of aluminum were combined with other design features to produce the lightest complete U.S. production engine of the time, approximately 300 pounds without flywheel or clutch. Further weight reduction occurred in 1965 through using an alternator in place of the generator. The lightweight engine concept did not detract from durability. Although the usual minor problems occurred, GM solidly backed these up with satisfying warranty service. The design engineers also paid attention and many running changes occurred over the years. And, as problems were designed out, parts interchangeability was retained wherever possible. In many instances, the later parts fit the '60 models, as detailed throughout this book. Not only did the engineers provide durability, their initial 80 HP (SAE) engine has proved capable of far greater performance. Race-modified Corvair engines, non-supercharged, are now exceeding 200 HP with good reliability!

Model Year Changes

This section includes basic body-style availability, engine changes, chassis changes, and indications of the more-important options. Many added details are in specific chapters of this book, of course.

1960—The initial-year offering included 4-door sedans and 2-door club coupes, with a choice of two engines and two transmissions. Both engines used a 2.6-inch stroke with 3-⅜-inch bore to get 140 cubic inches displacement, with the 80 HP using 8:1 compression ratio; the 95 HP 9:1, and a higher lift, longer-overlap camshaft. Powerglide and a 3-speed aluminum-cased transmission were supplied. All had 3.55 axle ratios, with an optional 3.89 available. 1960 mid-year changes included minor revisions to carburetors and a deeper groove to the idler pulley. The original two-stroke emergency-brake assembly was replaced with the single-stroke configuration as a factory-authorized change. In February 1960 the single-row ball-nut steering assembly was changed to a double-row configuration. Many parts are unique to the 1960 model. In some cases, newer items can be retrofitted to the 1960 models.

Ghosted view of 1960 engine to reveal finning on head, blower, and sparkplug location. Elimination of linkages for blower inlet control greatly cleaned up appearance of later engines.

The aluminum three-speed transmission case above was only used on early 1960 production.

The basic body structure of the Corvair remained the same through 1964, the easiest way to tell the years apart is by the front-end molding. Although items may have been interchanged over the years, each year had a distinctive front-end molding.

1961—A 4-door station wagon, Rampside pickup, Corvan, Greenbrier bus, and sports wagon were added to the body types. The nose of the car changed from a concave to a convex shape. The standard Turbo Air engine stayed at 80 HP and 8:1 c.r., but displacement was upped to 145 inches. The same displacement, a special cam, and 9:1 c.r. gave 98 HP to the Super Turbo Air high-performance optional package. A 4-speed transmission was added to the gearbox lineup, with revised ratios in the 3-speed which now resided in a cast-iron instead of an aluminum case. 3.27:1 became the standard car axle ratio with 3.89 and 3.55 continuing as options. Significant internal and external engine changes were made. Most obvious was elimination of the inlet control on the cooling fan with a change to revised shrouds with damper doors at the lower rear of the engine. This change was at least partially dictated by the needs of the new direct-air heater which was added as an option. Ducts added to the shrouds were designed to direct a small amount of warmed recirculated air over the engine at all times. Cylinder-fin stud-pass-thru holes were eliminated, thereby eliminating an advantage offered by the '60 cylinders: hoop-like fins that surrounded the cylinders to minimize barrel distortion. Engine overheat detection was changed from an oil-temperature measurement to a cylinder head snap switch adjusted to light the overheat warning at temperatures above 450° F. Oil slingers at the front and rear of the crankshaft were eliminated, a rear slinger came back on '64s. Combustion-chamber volumes were increased to retain the same compression ratios with the increased displacement provided by the 3-7⁄16-inch bore. In the carburetion department, external bowl vents were eliminated in favor of vents in the air-inlet horn. Manual chokes on each carburetor replaced the

1962–63 engine which showed few changes other than the use of automatic chokes on each carburetor.

1960's single automatic choke. Separate air cleaners with foam elements were introduced. The vacuum balance tube between the manifolds was reduced to ⅜-inch diameter. Nonmetallic resistance-core sparkplug wires replaced the copper ones previously supplied. A mid-year change extended cooling fan blower vanes to the hub center for reduced noise and improved air-delivery characteristics at high RPM.

The original gasoline heater was kept in the option column. The new direct-air heater (previously mentioned) was added as an alternate option. A two-speed windshield wiper was introduced along with windshield washers. Both were optional.

The 1961 Super Turbo Air engine, externally identified by a copper-colored oiler on the distributor, can also be identified by the engine application suffix and by larger valve springs and retainers installed into larger recesses in the heads. Compression ratio was 9:1 and a revised cam gave more overlap with 0.382 lift. Minor modifications were made to the exhaust system and the clutch-disc diameter was reduced to 8 inches to ease high-speed shifting.

1962—Basic body styles remained unchanged through mid-year. A convertible body was added in April, together with the introduction of another important option: an exhaust-driven supercharger (turbocharger). The addition of the convertible to the production line spelled the end of the station wagons. Transmissions and rear-axle ratios, with the addition of a limited slip Positraction option, remained static. The 98 HP Super Turbo Air was upped to 102 HP, Powerglide models with standard engines received a 4 HP improvement to 84 HP by using the 9:1 heads on the 80 HP engine.

In April of 1962 the 150 HP Turbo-supercharged Spyder was added, together with the desirable options of heavy-duty suspension and metallic brakes. The suspension change basically consisted of revised spring rates, a ⅝-inch-diameter front stabilizer (anti-sway) bar similar to that which had been previously available as an EMPI accessory, and stock-size rear shock absorbers with modified valving. Engine changes in the Spyder are so extensive that they have been covered in the "Spyders and Other Turbo-charged Corvairs" chapter.

The 80, 84, and 102 HP engines got automatic chokes on each of their two carburetors, and the Spyders were similarly equipped on their single side-draft Carter carburetor. A new distributor with built-in bearing was added to eliminate service difficulties experienced with shaft seizures in the earlier units. Revised "take-apart" engine shrouds provided much easier access to the items under the cylinder heads, i.e., the choke thermostats, oil pressure and engine-temperature senders, and the exhaust manifolds. Beginning in mid-1961 and carried into 1962, 9:1 cylinder heads had revised chambers with filled-in areas around the valves and sparkplug bosses. A plastic fan was to have been introduced in 1962, but it was withdrawn from production. The direct-air heater was made standard on all models. A heater delete option was still offered.

1963—Body styles were reduced by one with the earlier withdrawal of the station wagon. Transmission and rear-axle availability was the same as 1962. Options of metallic brakes, heavy-duty suspension and Positraction were continued. Minor engine changes included positive crankcase ventilation on all models, revised cam contours for quieter operation, and improved valves. Exhaust valves, except Corvair 95s, had four circular grooves in the stem for the keys (keepers) that help to impart rotation to the valves. The exhaust system used aluminized components for longer life through rust prevention. The original 4-plate "folded fin" oil cooler was replaced. A 3-plate was used in all except Spyder and air-conditioned cars, which received 8-plate coolers. An improved oil-pressure switch was introduced and relocated to the top of the oil-filter adapter. Self-adjusting brakes were added to all models. Spyder-engine changes were minor and are discussed in that chapter. Standard transmission equipped Forward Control models received a mid-year change to the shift linkage, giving much improved gear selection. 1963 can be said to be a "carryover" design from 1962 in nearly all respects.

1964—This was the year of "changes that count." Body styles remained unchanged except for the addition of a new engine-air-exhaust grille at the rear of passenger-car versions, and full wheelwell trim on the Monzas and Spyders. The Spyder now had its own model designation: it was formerly an option on a Monza coupe or convertible. Axle ratios were reduced to two offerings: 3.55 and 3.27, with the highly desirable (for acceleration) 3.89 gear banished to replacement-part availability only. Metallic-brake and heavy duty suspension options were also discontinued, but 40 radial cooling fins were added to the rear-brake drums on all models. Both of the transmissions were revised with the 4-speed getting beefier synchros and closer ratios, and the 3-speed returning to ratios as originally supplied in 1960. Input shafts were strengthened on both models of transmissions.

Handling got a lot of attention with the introduction of a ¾-inch-diameter front stabilizer bar as standard, complementing rear-suspension changes. The aluminum steering-gear box was dropped in favor of one cast from malleable iron. A one-piece steering shaft extended from the box all the way to the steering wheel. Rear suspension changes included reduced coil spring rates and the addition of a transverse leaf spring similar in appearance to that invented by EMPI and supplied for several years previous as an accessory for Corvairs and VWs. Some of the high-performance Super 90 and Carrera Porsches imported into the U. S. had this type of transverse spring. It reduces the roll couple and generally adds additional understeer to the suspension characteristics. Also at the rear of the car, they added a new differential dipstick, and new sealed-design U-joints at each rear axle pivot. Gasoline heaters were abandoned. Extensive engine changes are mostly internal and can be enumerated as follows:

Alloy-steel crankshaft with 2.94 (2-15/16 inch) stroke. This provides 13%

1963 coupe coming off of the production line showed continued use of "hip-line" styling which was copied by several European firms, including Fiat, NSU, Renault, and Simca. This body was given a label of classic styling by *Car Life.*

Tuned "snout" central air cleaner and harmonic balancer distinguished the 1964 from previous models. Stroke increase of 0.340 added 19 cubic inches to give 164-inch displacement. 110 HP High-Performance engine shown here.

displacement increase to 164 cubic inches. A shortened flywheel hub is combined with a revised flywheel-housing-seal location (0.070" rearward) to allow room for an improved bent-finger clutch and stepped flywheel. See "Clutch and Flywheel" chapter. A harmonic balancer was provided on all engines except basic 95 HP three-speed. It cancels vibrations created by the longer stroke, and increased power. Installation of the balancer required rear engine mount changes.

Cylinder-head chambers were redesigned to give minor compression ratio increase with the enlarged displacement. Valves were improved, and the large-diameter valve springs were used in all models. Pistons were modified for the increased stroke (pin height reduced), flat sides of the pin bosses were relieved for rod clearance, and piston rings were relocated (and redesigned). Cylinder spigots were notched and adjacent crankcase surfaces recessed to clear the increased swing of the crank and rods. Bore remained at 3-7/16 inches. More radical camshafts were employed, with repositioned lobes to clear the crank and rods. Spyder-type connecting rods were used with improved bearings in all models. Connecting rod caps were redesigned with narrowed ribs to clear the cam lobes. Oil coolers were again changed, with the '63 Spyder's 8-plate used on all 95 HP and 110 HP engines, and a 12-plate job created for the Forward Control, Spyder, and air-conditioned models. Air cleaners were changed, except on air-conditioned models, to a single centrally located oil-wetted paper filter similar to those used on the '60 models. Rocker cover mountings and gaskets were improved to help stop oil leaks. Ignition timing and carburetor calibration were changed to mate with the increased displacement. Clutch pressures were increased and Spyders returned to a 9-inch-diameter clutch disc. A revised clutch design with actuating-component changes included a shorter throwout bearing. A 35-amp generator became standard and complicated servicing, as it is unique to the 1964. The generator drive pulley and fan must be removed to take the generator off of the motor.

1965—Skin-deep changes are to be expected when a car gets a major face lift. But, the chassis and engine changes for '65 turned out to be far more important than the styling—even though the styling did tend to overshadow the technical improvements. Engines kept their 164-cubic-inch displacement, with the 95 and 110 HP engines essentially the same as '64s except for the addition of an alternator. High-performance (140 HP) and turbocharged (180 HP) engines took real jumps, with big-valve heads and four carburetors for the 140, and an improved intake and exhaust system accounting for 30 HP in the high-performance turbo job. Roughly the same boost was obtained for the turbo motors through drastic

1965 110 HP engine has only two carburetors and small-valve heads, shares cam with 140 HP engine.

Perhaps optimistically rated at 180 maximum horsepower, the late turbo motor of 1965/1966 was a great highway flier. When boost came in, there was enough horsepower to run away and hide from a 140.

The smog pump on the Corvair motor severely shrouded the distributor. There is one under there, somewhere.

To equip the 4-carburetor motor with air conditioning, the compressor was moved far over to the right side of the engine.

internal improvements to the turbo-supercharger and the addition of a larger-venturied carburetor. Corsas have a 270-degree sweep tachometer, manifold-pressure/vacuum gauge, cylinder-head-temperature gauge, trip odometer, and an electric clock with a sweep second hand—thus, a built-in rally pack. The 12-plate oil coolers are used on both Corsa models, and on the 140 HP engine when it is used in other series. All 1965 engine cylinders have wider head gasket sealing surfaces and the heads are bored out to accept the larger tops of the cylinders. Camshafts are identical to '64 offerings. Carbs of the non-supercharged models were modified to include a high-speed enrichment system. The suspension department had the most important changes. These converted the Corvair into the finest handling stock U.S. car on the road. The transverse-link rear suspension used on 1963 and later Corvettes was borrowed for the Corvair. Rear wheels stay nearly vertical in bounce/jounce and cornering situations—no more "jacking" of the rear end. Rear roll center was drastically lowered from the 13-inch position on '60–'64s to about 5 inches. Coil springs were retained throughout. Front sway bar diameter was increased from the $\frac{3}{4}$-inch used in '64 to $^{13}/_{16}$ to compensate for the reduced spring rates. Front/rear tread was increased to 55/57.2 inches and the 108-inch wheelbase was continued. Brake diameter and width increases added 43 square inches of available lining swept area (169 total). Driver comfort got attention with an optional telescoping steering column (shared with the Corvette) and wood-like rimmed steering wheel. There were a couple of mid-1965 model year running changes. The one-piece steering shaft, introduced for the 1964 models, was replaced with a two-piece steering shaft, equipped with a safety-enhancing stamped steel aligning coupler on the front side of the firewall. The throttle linkage for the 140 HP motor was replaced with a more forgiving spring-loaded link to the secondary carburetors. The only 1965 Forward Controls, Greenbriers, were dropped early in the 1965 model year, replaced by a conventional Chevy van.

1966—A valence panel added at the front not only concealed the suspension but added other advantages: higher top speed, increased stability, and improved gas mileage—and it fits the '65s, too. Primary mechanical changes were in the transmissions, both the 3-speed and the 4-speed now shared the larger Chevrolet's all-synchro construction. All gears are in constant mesh and sliding clutches lock each gear to the mainshaft as selected. Higher capacity synchronizers are used for all forward speeds. Wider gear design, the absence of any sliding gear action, shielded bearings, and a magnetic chip collector all increased transmission durability—long

Late model convertibles are often seen at concours events.

a Corvair sore spot. The differential dipstick was gone. Positraction (limited-slip differential) and a heavy-duty suspension and steering package were offered as options. The steering/suspension package was RPO Z17 which reduced lock-to-lock steering to three turns and provided flatter cornering. This option included a faster ratio steering gear and steering arms slightly shorter than standard. The engine line was unchanged from '65, except in California, where smog pumps were added to all engines except air-conditioned models and turbos. 1966s sold in California with the Air Injection Reaction (A.I.R.) systems were equipped with modified distributor curves.

1967—This was a "do-nothing" year for Corvair which many people felt was GM's "kiss of death" to close out the line, because the new Camaro was the performance leader. Corsa models were discontinued in 1967, leaving only the Monza and the Corvair 500. Although the performance-option 140 HP engines were not offered at introduction time, the 140 HP job was made available as a COPO (Central Office Production Option) bearing COPO 9551B as its designation. Any dealer could order the 140 HP option for any customer, the turbo, however was gone for good. The special-purpose suspension option was RPO F41, Positraction was G81, and fast steering RPO N44 brought you the 16:1 steering. New thin shell buckets appeared, courtesy of the new Camaro. GM had committed to continue to meet all smog and safety requirements, but that was about all of their efforts. 1967 Corvairs had a number of items added to comply with the federal safety and air-pollution requirements. Corvairs, and all other Chevrolets, received this long list of add-ons: front-seat shoulder belt anchors in coupes (actually added in mid-'66), a truly padded instrument panel, padded sun visors, four-way hazard warning flasher, dual master cylinder brake system with warning light, dual-speed windshield wipers, windshield washer, backup lights, outside rearview mirror, pushbutton seatbelt buckles in front/rear passenger locations, passenger-guard door locks, folding seat latches, day/night mirror with shatter-resistant vinyl-edged glass and breakaway support, lane-change warning lights in turn signals, corrosion-resistant brake lines, energy-absorbing steering wheel and column, smooth contoured knobs and levers on panel, soft low-profile window-control knobs and coat hooks, front seatbelt retractors, safety door hinges and latches, thick-laminate windshield, reduced-glare on instrument panel and windshield wiper arms and blades. This was the last year for four-door sedans. Air conditioning was no longer offered, neither the factory nor dealer installation, although the 1966 dealer kit was an easy installation. Again, all California cars had Air Injection Reactors (smog pumps). Chevrolet redesigned the system in mid-1967, and that system design was used through the end of the Corvair run.

1968—Astonishing many, GM continued Corvair production in 1968. There are rumors that the lawyers in some of the lawsuits advised GM that it would have been unwise to discontinue the car while still defending the car in court. For whatever reason, in 1968, GM again updated the car to meet revised safety and smog-control standards. All engines now came equipped with smog pumps, a smog pump equipped 140 HP motor was not a joy to work on. About the only external change to the 1968 Corvair was the appearance of new side marker lights both front and rear. Three models were offered: Monza sport coupe and convertible and the Corvair 500 sport coupe, the 4-doors were gone. 95, 110, and 140 HP engines were offered, all with any of three transmissions: 3- and 4-speed and Powerglide. New features included a

Chevy was testing V8 engines in the back of Corvairs in the early '60s. This stretched '65 housed a Cast Iron in the back. No notes on how well it handled, just a GM photo.

Experimental GM engine for Astro 1 coupe on page 161 demonstrates what might have been done with the Corvair basic engine. Power was 240 HP at 7,200 RPM: 1.4 HP per cubic inch. 176-cubic-inch displacement required only ⅛-inch overbore with stock 2.94-inch stroke. Light-alloy cylinder blocks were cast in threes with cooling passages between steel-lined bores. Two three-throat downdraft carbs were similar to those in Chapter 3, *Carburetion and Fuel Systems*.

Single-overhead camshafts were driven by Gilmer-type toothed timing belts from the crankshaft, which also mounted a double-sheave pulley for driving the three centrifugal cooling fans and the alternator. Combustion chambers were nearly hemispherical, with inclined valves actuated similarly to the OHC Pontiac engine.

refined instrument panel with recessed knobs to meet safety requirements, new interior trim styling, seat belts for all passenger positions, shoulder belts in coupes, padded windshield pillars, armrests shielding door handles, and a new vapor-return line in the fuel system for improved hot start and idling. Door hinges were revised again and are unique to the '68–'69 models.

1969—This was, after all, the last model year. There were some minor changes to the front brake lines and wheel cylinders. Changes were minimal—as you might imagine. The transmissions, engines, and body styles were identical to 1968, although a new-design clutch cable was introduced. All 1968 features were continued, and the 1969 announcement from Chevrolet listed four new features: new interior trim styling and colors, more comfortable bucket seats in Monza models, two front seat head restraints, and a new key design. All that really changed was build quality. All during the 1969 production "run," the handwriting was on the wall about the end of the car. The cars were produced at a very slow rate. Because the 1969s were the last, many cars at the end of the run were stored away by dealers and others wanting to have a "collectible." Many of those have resurfaced since, some with very low miles. By May 1969 Chevy had held on long enough and May 14 saw the last Corvair roll off the line, a gold Monza coupe Serial #6000, current whereabouts unknown.

U.S. PRODUCTION BY YEAR AND MODEL

	1960	*1961*	*1962*	*1963*	*1964*	*1965*	*1966*	*1967*	*1968*	*1969*
Horsepower	**80, 95**	**80, 95, 98**	**80, 84, 102, 150**	**80, 84, 102, 150**	**95, 110, 150**	**95, 110, 140, 180**	**95, 110, 140, 180**	**95, 110, 140**	**95, 110, 140**	**95, 110, 140**
Models										
Rampside		10,787	4,102	2,046	851					
Loadside		2,475	369							
Corvan		15,806	13,491	11,161	8,147					
Greenbrier		18,489	18,007	13,761	6,201	1,528				
Total U.S. Truck Production		47,557	35,969	26,968	15,199	1,528				
500 4-door Sedan	47,683	18,752				17,560	8,779	2,959		
500 2-door Coupe	14,628	16,857	16,245	16,680	22,968	36,754	24,045	9,257	7,206	2,762
500 Wagon		5,591								
700 4-door Sedan	139,208	51,948	35,368	20,684	16,295					
700 2-door Coupe	36,562	24,786	18,474	12,378						
700 Wagon		20,451	3,716							
Monza 4-door Sedan		33,745	48,059	31,120	21,926	37,157	12,497	3,157		
Monza 2-door Coupe	11,926	109,945	144,844	117,917	88,440	88,953	37,605	9,771	6,807	2,717
Monza Convertible			13,995	36,693	31,045	26,466	10,345	2,109	1,386	521
Monza Wagon			2,362							
Spyder 2-door Coupe			6,894	11,627	6,480					
Spyder Convertible			2,574	7,472	4,761					
Corsa 2-door Coupe						20,285	7,330			
Corsa Convertible						8,353	3,142			
Total U.S. Car Production	250,007	282,075	292,531	254,571	191,915	235,528	103,743	27,253	15,399	6,000
Total U.S. Production	250,007	329,632	328,500	281,539	207,114	237,056	103,743	27,253	15,399	6,000
Total Canadian Car Production	3,261	7,739	7,505	6,874	7,369	10,036	6,137			
CKD (Completely Knocked Down) for Export						696	432			

Sources: GM Heritage Center; *The Corvair Decade,* by Tony Fiore, published by Corvair Society of America; Corvaircentral.com; oldride.com

2

Measuring Horsepower, Performance Theory, and Tradeoffs

Before setting about to improve the performance of any engine you should become well acquainted with the stock article. For that reason, most of the modification chapters in this book begin with a general description of the stock components. Let us take a good look at the factory data before proceeding further.

The basic measurements used in evaluating engine performance, torque, and horsepower tell us how much work can be done in a particular period of time. One horsepower will raise 33,000 pounds one foot in one minute, or 550 pounds one foot in one second. Measuring engine output requires an engine dynamometer or power-absorption brake. These devices are comparable to a giant torque wrench with a slipping fluid or magnetic clutch. Some of them use water as the medium for absorbing work, others use an electrical generator. An arm on one side of the device is connected to a scale for measuring torque, and the crankshaft is coupled to a tachometer for measuring engine speed. The load can be varied to permit the engine to show its maximum capabilities at any given speed. When the torque and speed are known, the HP can be computed from the formula:

$$HP = \frac{\text{Torque (lbs. ft.)} \times \text{RPM}}{5252}$$

From this formula you can see that raising either the Torque or the RPM will increase HP. Many engine modifications merely move the torque output to a higher RPM, which gives more HP at higher RPM, and at the same time reduces low-speed engine flexibility. Others, such as boring and stroking, add torque and HP proportional to the cylinder-displacement increase. The effects of various modifications on engine flexibility are extremely important, as will be discussed in other chapters.

Published specifications for the Corvair show both gross and net power and torque figures for many of the models and we can estimate the others. The auto manufacturers were not trying to "fool" you when they quoted the gross HP figures, as they were following a Society of Automotive Engineers (SAE) Test Code, which was standard in the industry at the time. However, many automobile owners are not aware of the significant differences between the net or installed HP and that obtained from hand-built engines used to establish the engineering specifications. GM engineering sheets describe the methods used to establish Corvair data as follows (direct quote):

"The engine performance curves represent full throttle performance as obtained from dynamometer test data corrected to standard barometric pressure 29.92 inches of mercury and standard temperature of 60° F.

"GROSS POWER and TORQUE were obtained in a regular

1963 Spyder engine on test at General Motors Technical Center, Warren, Michigan. Engine is shown coupled to dynamometer which is automatically programmed to apply varying loads at specific speeds for life testing.

dynamometer test with the dynamometer exhaust system, no fan, generator not charging, optimum spark advance, and optimum fuel setting. NET POWER and TORQUE were obtained from a dynamometer test simulating actual operating conditions when the engine is in its vehicle, except the generator is not charging."

In the mid-1970s both the SAE and the manufacturers changed their reporting requirements and now SAE net horsepower is the allowed advertised number. This means you can't relate today's advertised horsepower to numbers from the 1960s. Net power is our main interest. When you look at the following listings it will become obvious why many engine builders expend a great deal of time and money in getting their engines to produce the advertised HP output. Believe me, any car producing its advertised HP at the flywheel will run away and hide from its puny showroom brothers who can only whistle up 70% of their advertised steam. Corvairs show almost 80% of advertised (gross) power, which may be due to the fact that they do not have to turn both a fan and water pump as do their water-cooled cousins. Here are the published statistics for 1960–65 Corvair torque and horsepower. Asterisked (*) figures are author's estimates where data were not included in the specification sheets. Note that the standard engine remained

HP & Torque Graphs

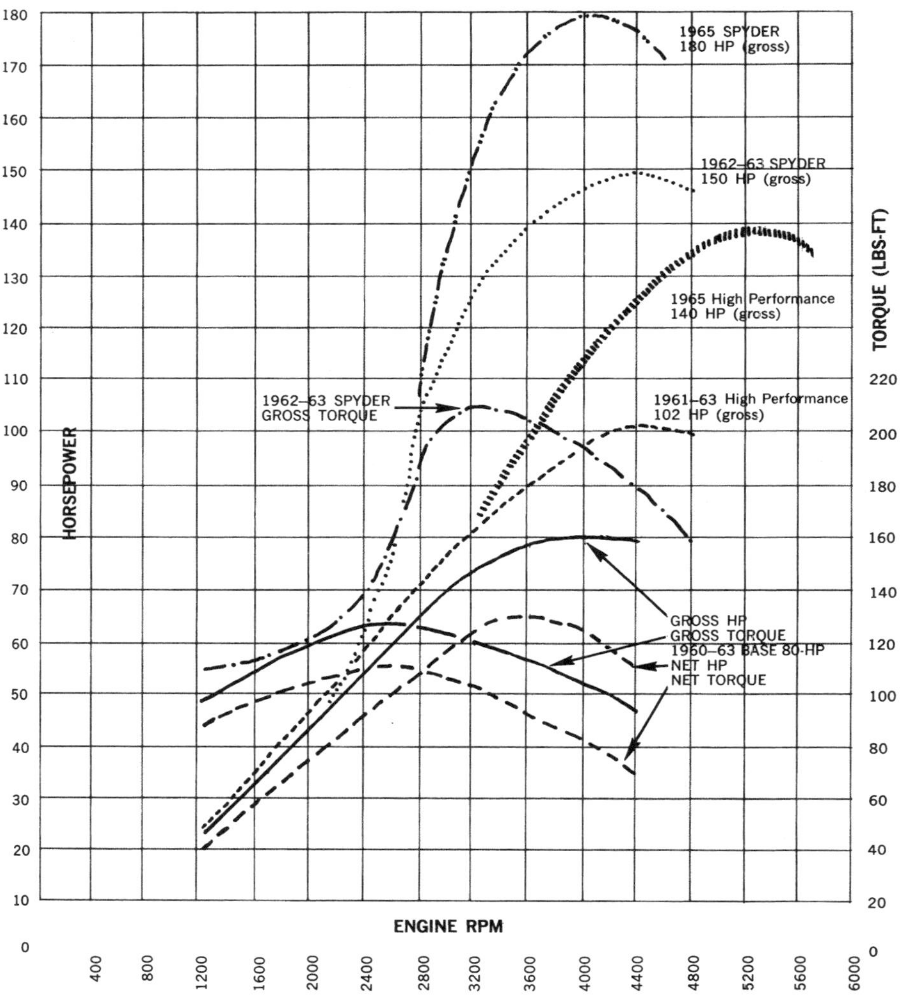

HORSEPOWER: All measurements in HP at RPM	Standard	Powerglide	High Performance	Spyder
1960 Gross	80 @4400	same as	95 @4800	not
Net	65 @3600	standard	76 @4800*	offered
1961 Gross	same as	same as	98 @4600	not
Net	above	standard	79 @4600*	offered
1962 Gross	same as	84 @4400	102 @4400	150 @4400
Net	above	68 @3600	82 @4400	120 @4400*
1963 Gross	same as	same as	same as	same as
Net	above	above	above	above
1964 Gross	95 @3600	same as	110 @4400	150 @4000
Net	78 @3600	standard	90 @4400*	132 @4000*
1965 Gross	95 at 3600	110 at 4400**	140 at 5200	180 at 4000
Net	78 at 3600	90 at 4400*	115 at 5200*	148 at 4000*

TORQUE: All measurements in lbs. feet at RPM	Standard	Powerglide	High Performance	Spyder
1960 Gross	128 @2300	same as	125 @2800	not
Net	118 @2200	standard	117 @2800*	offered
1961 Gross	same as	same as	132 @28-3000	not
Net	above	standard	122 @28-3000*	offered
1962 Gross	same as	130 @2300	134 @28-3000	210 @32-3400
Net	above	120 @2200	123 @28-3000*	190 @32-3400*
1963 Gross	same as	same as	same as	same as
Net	above	above	above	above
1964 Gross	154 @2400	same as	160 @2800	232 @3200
Net	140 @2400	standard	146 @2800*	210 @3200*
1965 Gross	154 at 2400	160 at 2800**	160 at 3600	265 at 3200
Net	140 at 2400	146 at 2800*	146 at 3600*	240 at 3200*

*Estimated as no authorized GM data available.
**Optional engine supplied with all transmissions (RPO-L62)

Bill Fisher's engine featured big-valve heads which provided 10.5:1 c.r. SS-H-30 camshaft and valve springs are made by Racer Brown. Manifold is center four-barrel IECO job. AFB Carter shown was replaced with a Holley for better resistance to turn cut-out. Oil pan, rocker covers, oil pump and valves are OTTO Parts. Pushrod tubes are baffled. Displacement of 176 cubic inches is provided by using '64 Spyder crank ('61 cases) with Badger 3-9/16 forged pistons. Crown adapter parts fit the engine into a Porsche roadster.

unchanged in advertised output for the four-year period 1960–63. Also notice the drastic improvement provided by the larger 1964 engines. Due to tolerance "stack up" it is doubtful that a production engine would measure up to these figures.

There are three basic efficiencies to consider when improving engine performance: breathing, or consuming maximum amounts of fuel/air mixture; extracting maximum energy from available fuel/air mixture; and reducing friction and pumping losses. There are important relationships between these volumetric, thermal, and mechanical efficiencies and they are not readily separable.

When thinking about breathing improvements, consider the engine as an inefficient air pump. Cause it to pump more mixture in and thrust additional exhaust out and you add horsepower. The simplest way to move more air through the pump is to make it bigger by boring and stroking. Other steps, usually more complicated, will include reducing inlet and exhaust-path restrictions. Everything cannot be changed to the optimum, but each of the following is usually improvable: air cleaners, carburetors, intake manifolds, intake and exhaust valves and ports, valve timing and lift, and the exhaust pipes, manifolds, and mufflers. As you improve the engine's breathing ability, it is enabled to turn higher RPMs, pumping still more air and giving higher output.

Getting the most energy out of the fuel being consumed as the mixture moves through the engine is largely related to two factors: compression ratio, and combustion-chamber efficiency. Improve Corvair's combustion chamber shape for better turbulence and you can employ higher compression ratios. Other simple shape changes induce better breathing by allowing the valves to function more efficiently. Higher compression ratios produce more power from a given amount of fuel, up to the point where detonation occurs. This point varies depending on the modifications that have been accomplished and the octane rating of the available fuels. Combustion efficiency is also aided by ensuring that a proper fuel/air mixture is supplied, and that appropriate sparkplugs fire it at the ideal moment for peak power and minimum heat losses. Additional energy can be gained from the fuel by polishing or coating the piston crowns combustion chambers, and the valve heads. Polishing or coatings reduce the heat loss into these parts and thereby ensures that more of the developed heat is available to push the piston. Corvairs were designed to utilize the available fuel from the 1960s. Although the 91–92 octane (R+M)/2 "premium" fuel of today is tightly controlled for emission reasons, it does not have the detonation resistance of the 100 octane (RON) fuel of the 1960s. Corvairs originally designed to run on regular gas, all of the engines with designated horsepower below 98, may today require the use of mid-grade gas, 89–90 octane (R+M)/2. If you are serious about racing a Corvair-powered vehicle, and day-to-day street use is not on your agenda, racing fuel, at very high octane ratings, is available, though quite expensive. There are several sources for 100 octane (R+M)/2 unleaded fuel, legal for street use, but it is still expensive. Because all Corvair motors used hardened valve seats in the aluminum heads, they do not need lead in the fuel as a valve seat protector, as do their cast-iron headed '60s counterparts. The lead may have anti-knock qualities, always a nice addition, but if the octane matches your requirement, ignore the leaded/unleaded controversy.

The last item is often overlooked or forgotten in engine assembly because it is so easy to overlook friction and pumping losses that are created in turning our "pump" and that reduce usable power availability. Improving the breathing reduces pumping losses by allowing the pump to turn more freely as it gulps the same amount of

The next time you are tempted to poke fun at Detroit automotive engineers, consider this 80 HP 1960 Corvair which was improved over a six-year period to produce as much as 180 HP—and there is room for more when it is needed. End ghosted/cutaway view provided by General Motors Public Relations Department shows relative positions of components.

air. Getting more power from a given amount of fuel also helps because friction is reduced to a lesser percentage of the total power being produced. Actual reduction of friction, however, is difficult to accomplish. Stock bearing clearances are close to the maximum that can be employed with reliability. Piston rings, combined with the piston, probably contribute the maximum friction found in the engine. The necessity for sealing the pistons is obvious, but research is constantly being accomplished to come up with new ring and piston designs to assist in reducing this friction. Major advances have been made in the 50 years since the Corvair's introduction. Piston clearances can be increased, but there is a limit, as excessive clearance causes piston "rock" and undue wear of the rings and grooves as their edges are rounded and pounded off. Even the oil system uses horsepower to force the oil through the confines of even the stock system. Still another friction source is misalignment of pistons and rods. This can steal power and generate unwanted heat. Careful assembly will reduce this item to a minimum value, of course. There are other friction sources, including the valve train, which cannot be minimized to any great extent. Balancing the rotating assembly further assists in reducing energy losses. It assures a smooth flow of power to the crankshaft flange, rather than a dissipation of energy into shaking motions.

Reading this book is going to give you all sorts of ideas about how to improve the performance and handling of your Corvair. But, before you begin spending time and money—STOP! Consider both the pros and cons of such. Can you afford to swap stock characteristics for the ones you think you need? For everything you get you must trade something in exchange . . . in addition to money! But with the perspective of 2013, we can see that Corvairs are not expensive cars. When Corvairs were new, and most people drove them as their only car, modifications had to be very carefully judged considering the impact on everyday driving and poor weather performance. Because the cars are now well into the

Modifications should be made to complement the major use of the vehicle. Remember mama when you start modifying your Corvair. 1965 Corvair Monza Sport Sedan represents one of the finest-handling 6-passenger sedans available from any automotive manufacturer. With minor improvements it can equal fine European sedans.

domain of "hobby" cars, more drastic changes can be considered, but the downsides must still be addressed.

These statements do not apply to careful tuning to get the last ounce of performance from the stock design. Nor do they apply to increasing the displacement by boring and stroking, to a point an almost trouble free-but expensive modification. Nearly all other changes couple disadvantages with the improvements. To make you fully aware of the things you can lose, I have carefully discussed the disadvantages of certain modifications. Do not overlook these statements as they were written with careful consideration.

In other words, added HP and handling are not always advantageous if you have to sacrifice too many desirable or required characteristics. As an example, F1 Grand Prix race cars have such highly developed engines that multi-speed gearboxes are needed for operation and the engine must be operated in a very narrow speed range to get any power at all. Were you to set out to extract the absolute maximum performance from your Corvair, this is the type of situation you could create. When the owners of some of the factory-special drag race cars decide to quit towing them and use them for transportation, considerable de-tuning is needed to make them drivable on the street. Be careful that you keep your modifications within the limits of the use expected of the car.

A special note of caution should be added for Powerglide owners. Be doubly careful what you do to the engine. It is absolutely essential to keep low speed torque with the Powerglide transmission. Failure to observe this caution could cause you to end up with a literally undrivable, horrible example type car. The manual gearbox overcomes or masks many of the problems coupled with HP increases but the Powerglide will not cover up for poor choices of engine characteristics or poorly selected modifications.

The same things can be said about handling modifications: consider the pros and the cons before spending time and money. You will often find that the more conservative, less-expensive modification provides the most usable driving fun! Careful analysis of where you want to go is always the first step in planning a trip.

In 1981, I retired an autocross car that had just won a regional championship. The '65 500 coupe had every conceivable mod which was allowed under the current "stock" rules. I had just acquired a '65 Corsa convertible less drive train. I swapped the complete engine and suspension from the race car into the convertible to drive on the street. I ended up with a poor riding, poor running, and twitchy handling street car. Over the next few years, I returned the convertible to stock springs, replaced the re-valved Koni shocks with stock type shocks, put on stock type carburetors (and linkage) and removed the metallic brake linings. Every time I changed something, it made the car more drivable and more pleasant to drive in daily use. Maybe Chevy was onto something there! Now, I'm not saying keep your car totally stock. The last forty years of car development have been too good to enthusiasts for us to ignore the fallout for our old Corvairs. But pick and choose your efforts and include time for thoroughly planning your destination and route.

Remember this paragraph when you are choosing a machine shop to perform some of your engine work. Choose a machine shop the same way you would choose a contractor to work on your house. You should be absolutely convinced of their quality and their dedication to performing the work you need. Prior good efforts for others, especially on Corvairs, are a good indicator, as well as the cleanliness and general look of their facility. Price should be considered, but consider the cost of doing it over, too.

MODIFICATION STAGES: ENGINE

ROAD	DUAL PURPOSE	ALL OUT
Carburetion		
Rebuild stock carbs (2 or 4). Add vent tubes. Consider bolt-on street EFI System. Modify stock YH carb for turn cut-out issues.	Bore venturis oversize. Center-mount 4-barrel with tuning. Performance port/type EFI for turbo, consider progressive throttle Weber DCOE.	Weber 3-barrels, 6-runner center-mount for carb or throttle body. Rules may dictate.
Ignition		
Stock distributor w/HD points or Pertronix Ignitor. Retain vacuum advance. Matching HD coil. Silicone plug wires. Hotter plugs. Brass-terminal distributor cap. Retain pressure retard on turbos.	Replace points with electronics (MSD). Match the voltage to the unit. Remove vacuum advance. Use 140 advance curve. Limit total advance to 36 degrees. Swap in alternate distributor such as 6-cylinder GM HEI or MSD. Consider electronic control of timing on turbos.	Crank trigger system or combine with EFI. Racing plug wires. Aftermarket high energy distributor with larger cap. Magneto is an option.
Don't use extended tip plugs if you have milled the heads. Inspect for piston interference.		
Exhaust		
Dual stock 140 mufflers. U-bends are available. Check tight left-side clearance on early models. Careful with noisy glasspacks. Big single helps clearance on early models.	Bolt-on headers—if lower shrouds are trimmed to fit. Flowmasters or other performance mufflers. Crossover ahead of mufflers helps noise issues. Regularly check exhaust donuts for sealing. No headers for turbos, wrap the crossover for quicker boost.	Slant-tubes in heads and flat flanges to custom headers. Mind the track dB ratings. Consider race mufflers with additional flange for SuperTrapp on tip or tips.
Cylinder Heads & Compression		
Careful valve job on stock heads. Seat valves to take advantage of full head diameter. Mill head to remove step and establish deck clearance to improve turbulence. 9:1 CR. Clean out all fins for better airflow/cooling through the head.	Use 140 heads. Check seats very carefully. 9.5:1 CR. Lighter valves. Prepare for larger springs. Some cleanup of carburetor pad holes for flow.	Remove manifolds for port clean up access. Consider "deep" valve seat replacements and new guides. Possible re-heat treat of head castings before machine work.
Cam & Valve Train		
Stock high-performance cam or replacement equivalent. Aftermarket springs to match. Inspect rocker tips for wear. Use proper locking nuts on studs.	Stock 140 HP cam is good to 6,000 RPM with correct springs. Isky 280 or Clark's 270 is a good street limit. Cam should be tailored to intake and exhaust flow. Take care during the installation to check all clearances.	Hotter cam is fine—Isky 280 to 310. Match springs. Consider roller rockers and stud girdles for longevity. Check rocker contact on valve tip. Retain set screw cam gear. Many hotter, dual-pattern cams are now available. Tailor cam to intake and exhaust.
Pistons, Cylinders & Bore Size		
Replacement pistons of Hypereutectic or forged construction. Stay at .030/.040 overbore for best cooling. Premium cast pistons are okay. Cast-iron rings or Total Seal rings are fine. Review piston skirt clearance per manufacturer's specs. Work with your machinist ahead of time!	New Clark's forged pistons on overbored stock cylinders. Full-finned Clark's cylinders are an option. Aftermarket piston/cylinder/ring kits, usually with Mahle pistons are a bargain.	Possible aluminum cylinders with forged pistons. Arias and others will do custom piston sets. Custom case modifications may be required for larger cylinders.
Crankshaft Stroke & Rods		
Start with late case and crank. Later rods are stronger. Try to stay at .020 or less undersize on journals. Always balance when any new parts are added.	Be sure to balance with the clutch/flywheel/dampener that you will use. New dampener recommended. ARP rod bolts/nuts should replace stock. If rods show any issues new forged rods are now available.	Best crank you can find. Zero-zero undersize is good. Magnaflux for cracks. Polish the journals. There may be aftermarket cranks out soon. Pin the output gear in place.
Lubrication Cooling & Breathing		
Clean the stock oil cooler before reuse. The stock '60–'62 folded fin or the late 12-plate cooler should be used. Add belt guides to stock shroud and pulley. Anodize or paint black the pan and rockers. Make sure all engine sheet metal is there and air-tight. Convert earlier models to magnesium fan.	Consider an external cooler. A simple one above the engine on the driver's side of the air inlet works. Plumbing off the stock cooler fittings will allow the fan belt to be changed without hose removal. A deeper oil pan and pickup will keep oil pressure up in turns. Use a spring loaded fan belt idler pulley.	Front-mounted cooler with temp-based bypass. Relocate filter as well. Larger oil pump is warranted. Use AN-10 lines to the front. Spring loaded idler pulley with a smaller or slower fan.

Carburetion and Fuel Systems

Most Corvair engines use two (or four) identical Rochester Model "H" series single-barrel down-draft carburetors. One is located on each intake manifold, separated from the manifold by an insulator block that also functions as a gasket. Choke arrangements have varied from a single automatic one on the air-cleaner inlet (1960), to manual operation (1961), to automatic chokes on each primary carburetor Model "HV" (1962 and later models). The Model "HV" carbs have the automatic choke mounted on the carburetor top. All 1960–61 and 140 secondary carbs are "H" series.

You also must remember that the Corvair was designed as an economy car, and sold as an inexpensive car. It had inexpensive carburetors, ones designed to last some finite number of years. We are way beyond the original design life for the standard "H" series Rochester, much like the remainder of the car. Things are starting to wear out on our vehicles that could never have been foreseen by Chevrolet.

Corvair's stock carburetion system is fine for economy and all-around uninspiring performance. The small-venturi carburetors (1.00-inch diameter, 0.785-square-inch area each) give good low-speed torque and fine engine flexibility, but restrict HP output at high engine speeds. Locating the carburetors close to the ports gives best-possible "snap" throttle response, but eliminates any possibility of using "ramming" for improved cylinder filling.

Some articles have pointed out that the stock carburetor positioning creates distribution problems. The truth is that it is correct from good engineering standards because it effects a reversal of flow between every pulse except when feeding the cylinders farthest from the carburetors (1 & 6) to straighten out what would otherwise be faulty distribution.

Simple changes to Corvair's well-engineered stock-carburetion system can achieve remarkable performance improvement. Careful tuning helps, but startling changes occur when four stock carburetors are used instead of two. Depending on your engine application, this should be the first performance modification that you make to your Corvair engine. When equipped with progressive linkage, this setup is superior for around-town operation, as well as all-out competition of road-race, drag-race, or Bonneville-type events. To get additional performance from further carburetion modifications can require exotic head changes, special manifolding, and expensive components. But we will talk about them just the same!

Center-mounted, multi-barrel carburetors exhibit faults that may outweigh their virtues when installed on Corvairs. A complete discussion is contained in the text.

Basics of Carburetion

The carburetor is a mixing and metering device, which combines air and gas automatically. It supplies these in the proper proportions to meet engine needs at varying speeds and loads. Although it performs a complex job, the Corvair Rochester carburetor is a

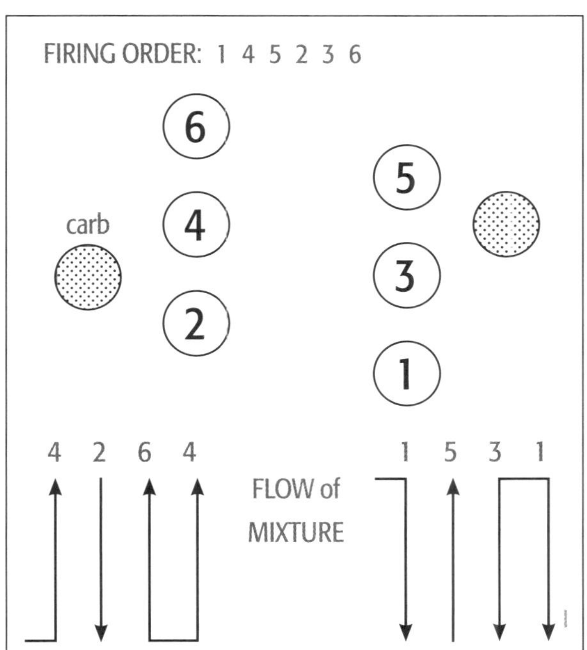

Chevy engineers must grow weary of reading about how they misplaced the carburetors on the Corvair and "fouled up the fuel mixture distribution." Regardless of self-styled experts' opinions to the contrary, Corvair carburetors are properly positioned according to known principles of flow theory. There is a reversal of flow at every point in the firing order (pulse through carburetors, in this instance) except when feeding the cylinders farthest from the carbs. By placing the carbs farther from the cylinders which are fed with a non-reversed pulse, the flow is evened-out to give correct distribution for this flat-six engine. It is hard to second-guess the factory engineers!

simple device because its initial and maintenance costs must be kept as low as possible.

The carburetor provides a restriction in the engine's air-inlet path to measure airflow and create a reduced-pressure area. This is called the Bernoulli principle. Jet openings placed in the reduced-pressure (venturi) area connect to a float bowl kept vented to atmospheric pressure. Although the jet is located slightly higher than the gasoline level in the float bowl, any decrease in pressure at the jet causes fuel to flow through it. As air is sucked through the carburetor by the piston strokes, fuel from the jet is added to the air stream in relation to the speed of the airflow. Fuel combining with the air increases more rapidly than does the weight of the air (mass). Thus, a jet that would provide the correct fuel/air mixture (approximately 15:1) at medium speeds would give a too-rich mixture at high RPM. Or, a jet correct for high speed would provide a too-lean mixture at lower speeds. For that reason, carburetors have compensating devices that include special jets, air bleeds, and emulsion tubes. These devices actually improve fuel atomization because they cause air to be mixed with the fuel ahead of the jet. Thus, when fuel flows into the venturi area, it is already partially mixed into an air/fuel emulsion. This emulsion is more readily atomized by the fast-moving airstream in the carburetor.

At all speeds above idle, the air passing through the venturi determines the vacuum present at the jet (pressure drop) to draw fuel into the air stream. If air speed is too slow, atomization is not as complete and fuel deposits on the manifold sides instead of continuing to the cylinders. Cold temperatures tend to exacerbate this problem. For this reason, adding carburetion capacity tends to cause flat spots at low speeds and to reduce engine flexibility. Increased venturi area requires a higher engine speed to keep air velocity sufficient to ensure good fuel atomization and distribution. If you quickly open (floorboard!) any multi-carburetor arrangement with large venturi area, the engine will misfire momentarily. This is because manifold vacuum rises almost to atmospheric, air velocity drops, and fuel drops out of suspension and onto the cylinder walls, or doesn't move at all. The mixture, now mostly air, arrives at the cylinder too lean to support combustion. Many auto enthusiasts think that quick throttle opening "floods" the engine with excess fuel. 'Tain't so!

Really sharp tuners recognize the basic facts about pressure differentials and the added complicating factors of air density and pressure. These fellows utilize accurate instruments to help tune their carburetors for championship performance. In general, as barometric pressure goes down, and/or temperature rises, your engine's horsepower decreases. You can partially compensate for such common atmospheric changes by changing jets. A normally aspirated car will lose power at higher elevations because there is just less oxygen to use to burn fuel. You can optimize mixture and/or spark for higher altitudes, but the output will be less.

The basic instrument and tuning charts showing you how to make the necessary compensating adjustments to your carburetion are available from several sources.

All Rochester H and HV models through the years are basically the same carburetor, with the exception of Spyder models. At least 11 such carburetors are listed (as of 1964) for replacement without including the Carter YH series side-draft carburetor used on the Turbocharged Spyder. Two more were introduced in 1965 to provide a primary with a high-speed enrichment circuit, and a secondary model for 140 HP 4 × 1 models. Spyders got a new Carter YH with larger venturi for its horsepower-boosting capability. Actually, many of these models differ only in minor lever and choke configuration changes. These are needed to cope with various linkage and space considerations of the passenger cars, trucks, and station wagons which these engines power. Smog pump equipped cars introduced several additional new models into the mix. Jet changes from model-to-model and year-to-year have actually been quite minor, not too surprising as the venturi size has never varied.

Production Carburetor No.	**7019111**	**7019117**
Used on Engine	Turbo-Air (80 H.P.)	Super Turbo-Air (98 H.P.)
With Transmissions	ALL	ALL
Throttle Bore	1-¼"	1-¼"
Main Venturi	1"	1"
Radial Venturi	.050	.050
Idel Needle Orifice	.046	.046
Idle Tube Restriction	.032	.030
Main Metering Jets	.050	.051
Main Well Vents	.045	.045
Cluster Top Bleed	.040	.040
Cluster Side Bleed	.040	.040
Pump Jets	.022	.022
Siphon Bleed	.059	.059
Cluster Channel Restriction	.033	.033
Main Well Insert	Yes	Yes
Adjustments		
Float Level	1-13/64"	1-13/64"
Float Drop	1-¾"	1-¾"

Table from Chevrolet Service News shows 1961 carburetor specifications. 7019111 superseded 7019100-101; 7019117 superseded 7019107.

1960—Model "H" Rochester carburetors were introduced on the 1960 Corvair. They were created especially for these engines. No power-enrichment system means that these carburetors run slightly lean at high speeds, requiring enlargement of the main jets for peak performance. Most early models were jetted lean for better fuel economy, remember the Corvair was an economy car, and Chevy was getting killed by the Falcon in the "Mileage Race." Main fuel-distributing cluster ends in four horizontal radial tubes instead of the more-common ring-type nozzle. This was done purposely to improve fuel atomization at low speeds. Do not look for choke mechanisms in these carburetors: the choke was in the air cleaner. Production changes in the 1960 carburetors included temporary use of a spring-loaded needle seat with dimpled float. This was later abandoned.

1961—Manual choke assemblies were added on each carburetor. A dash-mounted cable-pull knob allowed simultaneous choke operation. Choke cables and the linkage necessary to control both carburetors together caused many problems over the years. It was an interim solution until the revised automatic stove-chokes were ready. Fuel distribution cluster moved closer to point of maximum venturi restriction. External float-bowl vents were eliminated on most models. Float-bowl venting is by an air-balance tube inside of the air horn. If an external vent (1/16-inch hole) is found in the top of the air horn casting it will allow gasoline fumes to enter the heater system. (Close such holes with a 3⁄32-inch lead ball). This includes carburetors 7019100, 7019101, and 7019107.

On carburetors 7019100 (PG) and 7019101 (manual), engine surge sometimes occurred. Correction was accomplished by opening the main-metering jet to .050 inch on 80 HP or .051 inch on 98 HP and opening idle restriction from .030 inch to .032 inch This can be done by replacing the part from another carb or by drilling out the jets. Hold the drill in a pin-vise for doing such work, and be careful to keep the drill straight so that it does not drill oversize. Even if you drill the jets yourself you must add the main-well insert tube to each carburetor. This is Chevrolet Part No. 7009212. In addition, a matching venturi-cluster gasket No. 7020195 should be installed.

1962–1963—Automatic chokes were added to each carburetor, probably because of difficulties with functioning of the 1961's manual chokes. Slides on the '61 chokes were prone to stick, especially when not lubricated or operation was in very dusty areas. And, there is always the story about the elderly lady who kept complaining of poor economy—it turns out she always pulled out the choke because it made such a fine place to hang her purse!

The automatic choke on each carburetor consists of a thermostatic control unit mounted to the lower side of the cylinder head, linked directly to the carburetor choke-valve shaft; a vacuum diaphragm mounted on the air horn to provide start and run position of the choke valve, and a fast-idle cam to control engine speed during warm-up. Although you may be tempted to convert a '61 over from manual chokes to automatic, remember the hole through the cylinder head for the thermostatic choke link is not there in the '60–'61 heads, nor in most turbo heads. The stock '62-up chokes won't work with the early heads.

1964 and later (mostly)—Although basically similar, the carburetors were recalibrated to coincide with the increased airflow characteristics of the larger-displacement engines. An idle-air-bleed passage drilled through the base flange and into the throttle-bore wall improves "hot-soak" starting and warm-weather idling. The bleed is closed at all throttle positions beyond the highest step on the fast-idle cam. Sealing is by a small rubber plug mounted on a flat spring. Moving the linkage to the closed throttle position unseats the plug, opening the bleed passage. This vent may be a source of smelly fuel vapors if the hot-soak boils fuel in the carburetors and into the throats after shutoff.

A float-assist spring aids in stabilizing the float-bowl level. This was probably an alternative to remounting the carburetors to face float bowls forward. Part No. for the spring is 7028503, originally 15¢ each. One is required for each carburetor. The float must have the correct hole to use the spring. Extra bowl vents are standard on the '64s. The secondary carb on the 140 HP motor was basically a less-expensive version of the primary, if that was possible. Originally, no idle system was used on the secondaries, no choke, no lower vents, no accelerator pump. Some things were added back on later model secondaries. Some smog primary carbs have additional calibrated air bleeds in the main well so it's pretty easy to spot them.

The accompanying table from Chevrolet Service News, November 1963 (1964 carburetors) is shown here because the one in the Shop Manual is inaccurate. Carburetor No. 7024022 is used on 95 HP Powerglide models, whereas the 7024023 is used with synchromesh transmissions.

An excellent reference is Bob Helt's book *How to Identify and Rebuild Corvair Roadster H and HV Carburetors*, see Resources.

Improving Stock Carburetion

There are many reasons why Corvair owners want to improve the stock two-carburetor setup for utmost performance. Foremost of these reasons might be to contend in "stock" classes at either drag or autocross races. Some owners merely want to make sure that they get every last ounce of performance available from their stock engines. I certainly agree with either kind of thinking.

1964 CORVAIR AND CORVAIR "95" CARBURETORS

PROD. CARB. NO.	VEHICLE APPLICATION
7024023	5-7-900 Series w/95 hp bas engine (not w/Air Cond.
7024024	5-7-900 Series w/RPO L6 110 hp engine (not w/Ai Cond.)
7024025	All engines in Corvair "95" vehicles; all 5-7-900 Serie vehicles w/Air Cond.
3817245	Monza Spyder, Series 60 (Turbocharged)

SPECIFICATIONS	ROCHESTER "H"			CARTE "YH" (3692S
	7024022	7024023	7024024	381724
Float Level	1 1⁄16"			5⁄8"
Float Drop	1 3⁄4"			2 3⁄8"
Pump Rod	Index Line			—
Automatic Choke Setting	2 Turns Up from Free Entry in Lever			1 Notch Lean
Unloader	.312"			7⁄16"
Fast Idle Setting	.078"			—
Choke Piston Vacuum Break Adjustment	.180"-.195"			—
Main Metering Jet	.053"	.051"	.053"	.104"
Idle Speed Jet	.025"	.025"	.024"	.031"
Bowl Vents	3-Internal	2-Internal		1-Interna
Metering Rod (Size)	—	—	—	.078" / .046"
Pump Discharge Jet	Two @ .022"			2 @ .026"
Throttle Bore	1 1⁄4"			1 1⁄2"
Main Venturi	1"			1 5⁄16"

1965 HV-model specs are similar to those in this 1964 table except: float drop is 1-½ inch, main jet 0.051 or 0.052, idle 0.024, and two internal vents are used with one external vent

The H-series, in stock form, is prone to gas feed problems during cornering. It wasn't as much of a problem when relatively narrow bias-ply tires were the standard. But with newer sticky tires and better-constructed radials of the '80s and beyond, the stock carbs will give up the ghost. The positioning of the carb on the Corvair engine is the culprit. Under hard cornering, fuel flows away from the jet on one carb, leaning out that side while it allows extra fuel to build up on the other side. But the sloshing fuel can raise the floats on both carbs and, on a long sweeping turn such as a freeway on-ramp, you can run out of fuel in the carbs.

The fix is to allow more fuel to remain in the carb, and not let it dump down the throat. For street use, with normal tires, simple venting and careful float settings will cure most problems. The drawing at right shows how Doug Roe corrected turn cut-out problems on the Rochester H carbs. The original vent is closed with a lead ball. (This is done at the base underneath.) Two copper tubes are inserted so that they extend

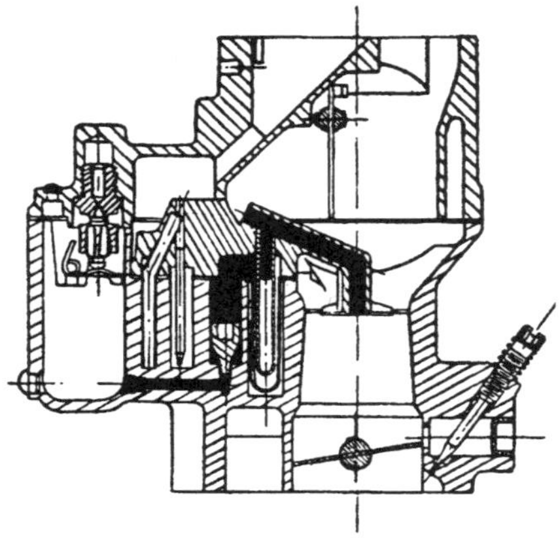

1965 Corvair carburetors are model HV which include a high-speed power enrichment system. A power-enrichment needle is located in a passage below and alongside of the main well. As venturi pressure drops on the main-well surface the needle raises off its seat to enrich the fuel mixture. Drawing courtesy Chevrolet Service News.

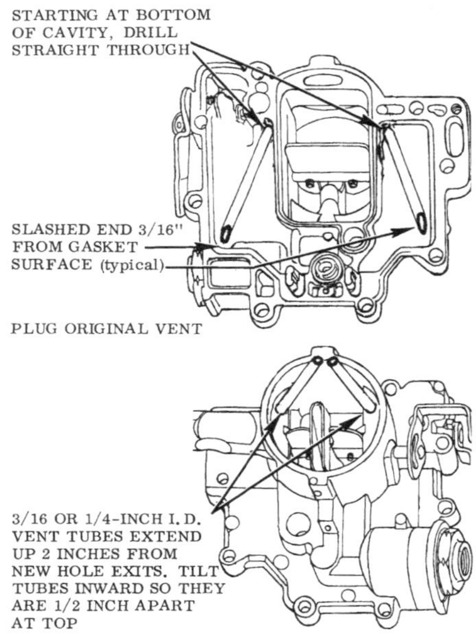

Doug Roe's recommended turn cut-out modifications ensure that one carb vent will always be open, regardless of the severity of cornering.

upward from the air horn opening approximately 2 inches, and tip inboard within ½ inch of each other. If a stock air cleaner tube is required, the copper tubes can be cut just about 1/2 inch above the air horn, just add 6 inches of rubber hose to each tube and run it up the air filter housing. Vertical height is what's needed, and this still looks stock. The "closed" gasket is used, and the float level must be just above production level specifications. This setup also cures stalling after long idling. Doug's original tubes dropped down into the float bowl, turned and extended to the corners of the area just above the gasket. The reason copper was used instead of the easier to find steel tubing, was that tight bend. If you just raise the fuel level in the carb by adjusting the float down (remember the adjustment is done with the carb top inverted) without adding the vent tubes described above, you will get flooding of the carb during cornering. Another method of fooling the fuel jet is to relocate the jet to another place in the carb, a place that will stay submerged in fuel under any cornering situations.

The multi-step procedure for relocating the jet involves adding a new feed passage to the main well, tapping the end of the passage to allow mounting of the jet, and plugging any old circuits feeding the well. The simplest way to add the circuit is to utilize the path that GM did when they added the high-speed enrichment circuit, but extend it all the way to the main well. (1) After the passage is countersunk (2) and tapped for the jet, (3 and 4) the newly expanded hole in the outside of the carburetor must be threaded for a ⅛-27 NPT pipe plug (5), many types of plugs are available. The high speed enrichment chamber hole under the cluster must be threaded and plugged with a 5⁄16 straight set-screw (6), and the original jet (7) must be removed and plugged with a ¼-28 set screw. Now the jet can be changed by removing the pipe plug and sliding in a straight-tip screwdriver.

Naturally, the first order of importance is bringing the carburetors up to the latest-year specifications, as shown in the charts. And, you can add the additional venting and float-torsion springs of the '64 models as well. The accompanying illustrations identify the jets and other parts. One of the first things many "mechanics" do to improve carburetor performance is to rip out the chokes. They are fooling themselves, of course, because there is usually so much area around the chokes that airflow is not aided by their removal. The tiny 1-inch venturi area is the object of our interest.

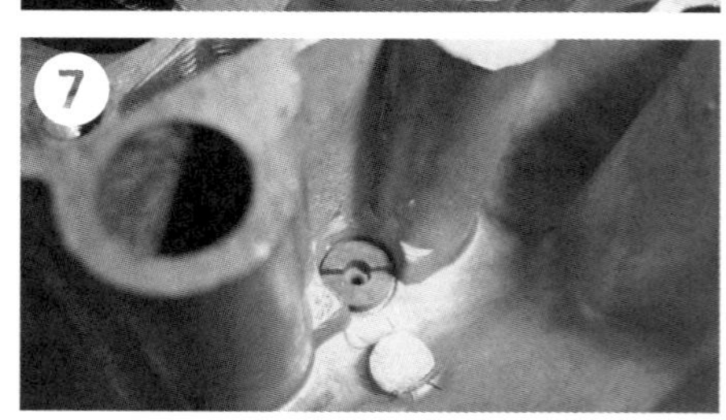

The fuel-distribution cluster obstructs flow through the venturi so streamline this part without drastically reducing its strength or changing the length of the four little arms.

When setting up the carburetors, back out the throttle-stop screws so the throttle plate fully closes. With a light shining from behind the carburetor, look into the air horn to observe whether the throttle plate is seating all the way around. Light around the plate edge indicates improper seating. Loosen the screws holding the plate into the shaft. But don't remove them, they are staked in place to prevent loss. The shaft should then hold the plate firmly, but not tightly. Then, while holding the throttle closed, tap lightly on the plate with the end of a screwdriver or drift pin to cause the plate to center into its bore. Tapping the shaft ends, to move it from side to side in the bore, also helps in some instances. Retighten the screws and again use the "light check" to determine whether you have established proper throttle-plate seating. If the throttle shaft hole in the carb base and/or the shaft is worn, there are a couple of options to fix it. The original shaft is plated brass and replacements are available in stainless steel. You will have to rework the throttle arm a bit to mount it on the replacement shaft. If the shaft hole in the carb body is not too badly worn, replacing the shaft may restore idle mixture control. There are carburetor remanufacturers around the country who will install bronze bushings in the carburetor base and ream them to size. If you really want a solid idle, this is the way to go. The O-ring/washer/spring treatment may help, but it can cause linkage problems, because the added friction tends to slow or prevent the throttle from returning to a fully closed position. Extra springs may help. If you do remove the throttle plate screws, be sure to restake them so that they won't loosen up and desert you. There is no good place for them to go.

If you do remove the chokes, say for racing, plug the shaft holes. This can be accomplished by tapping the holes 5⁄16 inch-18 and installing set screws. The set screws can be retained by center punching, or with Loctite. If you prefer, tap the holes and fill them with an aluminum-epoxy mixture such as Devcon F. Filling these holes forces the air cleaners to filter all the incoming air and simplifies using a carburetor-synchronizing gauge.

In general, original Corvair main jets were too small for maximum performance. This was to keep the best economy, of course. With the advent of oxygenated gasoline and lower octane ratings, a richer mixture is almost a necessity. You should consider using jets at least 0.052-inch diameter, from Rochester those would be referred to as #52 jets, many use #54s. Even larger jets have sometimes shown additional performance. Remember that anything done to improve the breathing of the engine may require enriching the mixture. Some of the changes that can affect carburetion in this manner include valve size, displacement, ignition timing, special camshaft, etc. Also, an efficient exhaust system plays an important role in the engine's overall breathing ability. Properly exhausted cylinders have less residual burnt charge and more intake charge can be handled, which may require mixture enrichment through the installation of larger main jets. Too rich a mixture uses extra gasoline, too lean a mixture can destroy engine components. Racers still drill jets, but the manufacturers don't recommend it, because they would rather sell you new ones. Oh yes, and they say the jets are sized on flow, not necessarily on drilled-hole size. All the different sizes are available for the H-series carb. The H series uses the same jet design as the Rochester 4GC and the Quadrajet. The threads are not the same as Holley jets.

Naturally, you will want to make sure that the carburetors are clean, carefully adjusted as to float level, and that the sintered-bronze filter element in the large inlet nut has been cleaned. Soak it in carburetor cleaner or a solvent such as acetone, and then blow it out from the reverse side with compressed air to remove any clogging contaminants. If you have any doubt about the condition of these little filters, replace them.

It is generally assumed that the smog (emission) valve, the positive crankcase ventilation system, can be dispensed with but its removal will drastically disturb the idle mixture on a carburetor system designed for use with it. Keep the system clean and it will not disrupt the carburetion or reduce performance in any way. Do not bore out the orifice in the late model PCV systems. If it seems plugged, clean it with a paper clip or tiny drill bit. If drilled oversize, it will present an idle air leak that will have you chasing your tail to adjust.

Another point of interest: the oil-soaked polyurethane filter elements (1960–63) do not always provide the same degree of restriction and can confuse your tuning efforts unless you always follow a set procedure for cleaning and reoiling. Their degree of restriction depends on how they are cleaned and reoiled. You may want to replace these with paper element filters for consistent performance. Most manufacturers make only paper replacements. There are aftermarket high performance filters available from K&N and Amsoil for the '64–'69 models.

Be sure that the insulator/gaskets are properly positioned on the head manifold openings before replacing the carburetors. There are thin gaskets available that can be placed above and below the insulator. I don't use them and I haven't found a real need for them. Improperly positioned gaskets will overhang into the inlet-air path, losing more HP than all of your careful tuning and expensive parts have gained.

An inexpensive, simple way to add performance is to install larger carburetors or bore out the venturis of the existing stock carburetors. A pair of 1-¼ inch venturi carburetors will provide a 56% increase in usable venturi area over the stock 1-inch venturis. Bores of this size will require installation of a sleeve in a bored-out carb. The stock venturis can be bored out to 1-⅛ inch diameter to increase total venturi area from 1.57 to 2.0 square inches, a 30% increase. Do not bore out to equal the throttle bore or performance will be impaired. Extend radial arms of fuel distribution cluster 1⁄16 inch. Enlarge main jets to 0.056 to 0.060 inch. Low speed torque will be slightly worsened, but at least 8 HP can be expected at the mid-range and top end. It works fine with Powerglides, also. The same modification can be applied to the 4 × 1 arrangement, too. Try boring the secondary carbs first on a 4 × 1 system. It may be all you need. The Rochester H series carburetors, as originally equipped, had a metal fuel inlet needle and a brass seat. Almost immediately a neoprene needle was supplied, and most kits still use them. Old age will catch up with neoprene, however, and the needle should be replaced during the rebuild. There are a number of newer design needle and seat replacements. The Grose-Jet uses a pair of round balls to seal the seat. They are hard to find, though, and have been out of production for years. Also, for any performance application, replace the bronze fuel inlet filters. They are your last hope of keeping crud out of the carbs.

Road Tests Provide Intake-System Comparison

This table was compiled from road-test figures published in *Car Life* and *Car & Driver* magazines. Although run in 1961–62 on the 102 HP Corvair, 1963 tests confirmed the earlier results. These data are probably quite accurate.

A–Stock 2 × 1, 1.57 square inches of venturi area: Rochester Hs

B–2-barrel carburetor on central manifold feeding both heads. Venturi area 1.88 square inches: Rochester 2 GC.

C–4-barrel carburetor on central manifold feeding both heads. Venturi area 3.44 square inches: Rochester 4 GC.

D–4 single-throat carburetors (4 × 1), two carburetors on each head. Venturi area 3.14 square inches: Rochester Hs.

	A	B	C	D
0–40 MPH	6.8 seconds	7.2 seconds	7.0 seconds	6.2 seconds
0–60 MPH	15.5	15.5	15.0	12.1
0–80 MPH	35.0	32.5	21.1	20.4
¼ mile	20.3	20.3	19.9	18 5

Racing Only Modifications

If you want to get serious about Rochester Carburetors for all-out performance, talk with your local Stinger racer. When Yenko went racing with the "factory" team, RST Racing, one of Yenko's engineers built the all-out carbs referred to as "Bill Hartley Webers" around the dealership. With a relocated jet, bored and sleeved throttle and venturi, (throttle at 1.5 inch, venturi at 1.25 inch) four of these on a ported 140 head easily out-flow the SCCA legal Webers. They don't have quite the mid-range of the Weber, but top end power is superior.

For several classes of competition, such as vintage road racing, the factory supplied Rochesters are usually required. There are several racing only modifications that can help racing performance. These are not recommended for street-driven motors. The chokes can be removed and the top of the carbs can be machined to remove choke restriction and "funnel" the air down. A lathe, with the proper fixture, can hold the top in place as the bore is trimmed. This should be done before adding any vent tubes. Beyond the bored out venturi mentioned above, the actual base of the carburetor can be bored out and a sleeve epoxyed in. This will allow 1.5 inch throttle plates to be installed, providing more air and fuel to the engine. This is not a simple procedure, as it involves re-boring the throttle shaft holes and adding back an idle circuit feed and idle/transfer ports. To utilize this modification, other changes must be incorporated, such as extending the tips of the cluster in the throat of the carburetor. Wild jetting will also be necessary, often well into the .070s, to provide enough fuel for the weakened airflow signal. Several all-out carbs have been built with jet-holders removable from the top of the carburetor, much like the DCOE/IDA Webers. Although not a power producer, this would make jet changes a much more pleasant proposition.

Four Carburetors for Corvairs? Four?

This is an impressive number of gasoline-mixing bowls, but there is universal agreement about the quantity and type to be used. The factory release of the 140 HP option in 1965 confirmed what the aftermarket had already known. Corvair enthusiasts and expert tuners alike say that this modification is the first step to take when urging more horses from the Corvair stable. Venturi area is doubled to a very useful total of 3.14 square inches. Placing the carbs close to the ports retains good snap-throttle response and ensures an improvement in fuel distribution over that provided by the stock arrangement. Carburetors discharge into a warmed manifold for good fuel vaporization. Easier starting is coupled with gasoline mileage on the order of 20 miles per gallon in around-town driving. Finally, the 4 × 1 gives improved power through most of the usable engine-speed range, providing better than-Spyder acceleration through 60 MPH.

How much more horsepower does the 4 × 1 setup give than the stock 2 × 1? That's one question not easily answered because Corvairs, except at General Motors, have not been widely tested on engine dynamometers. This is because the reverse-turning Corvair is not easily adapted to the average engine dyno. True, many data have been obtained on the rear-wheel (chassis) dynos, but these are not to be trusted. This is because owners of such equipment often "gimmick" them to save constant explanation to owners of expensive new cars who cannot appreciate the difference between advertised HP, actual HP, and rear-wheel HP—of which the latter is the smallest figure. Always be wary of chassis dyno readings! They are good comparison figures where the same dyno is used to make one change after another, especially on the same day. However, averaging rear-wheel HP figures from several dynos indicates as much as 15 HP improvement. A 14 MPH increase in terminal speed in the quarter-mile has also been reported. And, in passing, it should be noted that nearly all of GM's experimental cars were equipped with the 4 × 1 modification, with the exception of Spyders, of course.

Four progressively linked carbs are standard on '65-'67 140 HP engines. Because '65-'66s had no secondary idle systems, they often stuck shut. 1967 models have idle systems in all carbs, with no accelerator pump cup in secondary carbs. No. 7027026. Installing later carb cures the problem on '65–'66s.

Disadvantages are found in complex installation, carburetor synchronization, added cost, and slightly higher fuel consumption if the added performance is fully utilized. In addition, the after-market 4 × 1 kits did not provide for efficient air cleaners and special ones must have been fabricated. Air cleaners are essential to provide against fires due to starting "belch-backs," to reduce engine intake noise, and to reduce wear—regardless of how dust-free your area of operation may appear. 4 × 1 is not recommended for Powerglide-equipped Corvairs unless a progressive throttle linkage can be fitted.

Although there were two basic methods for mounting the two extra carburetors, kits that provide accessory flanges which mate with an extra hole cut into the top of the stock manifolds were most often used. These kits can be installed without the necessity for removing either the heads or the engine. The manifold is 0.200-inch thick, which is more than adequate for the installation. After cutting a hole into the manifold, the flanges are screwed on, using an epoxy-aluminum sealing compound to prevent any air leaks. Kits made by Ansen used a combination adapter that reoriented the carburetor and provided the added flange—all in a single casting for each head. Eelco/Thomas kits (they were the same) used star-shaped adapters atop the stock carburetor mounting flange, and separate castings for the added carburetors. Both kits aligned the carburetors with the throttle shafts parallel.

Back in the day, none of the manufacturers of Corvair four-carb kits provided a progressive linkage. Such linkages provide a definite improvement in low-speed performance, but they are expensive to make in small quantities Not many Corvair manifold kits were sold. Even so, it should be pointed out that progressive linkage for the 4 × 1 arrangement is the way to go, even for Corvairs being raced. Such linkages can make the setup suitable for use with Powerglides, too. You can make your own with Heim Ball Joints at all rotational joints. The difference in ease of synchronization and keeping a good idle with low spring tension must be experienced to be believed.

The second method for getting the 4 × 1 arrangement is to install a set of the 1965–67 140 HP heads. The 140 heads also offer the advantages of much larger valves and exhaust stacks. You can use stock complete primary carburetors (with idle circuits) at all four mounting pads. The '65–'66 stock setup uses secondary carburetors that have no idle circuits and can soon become

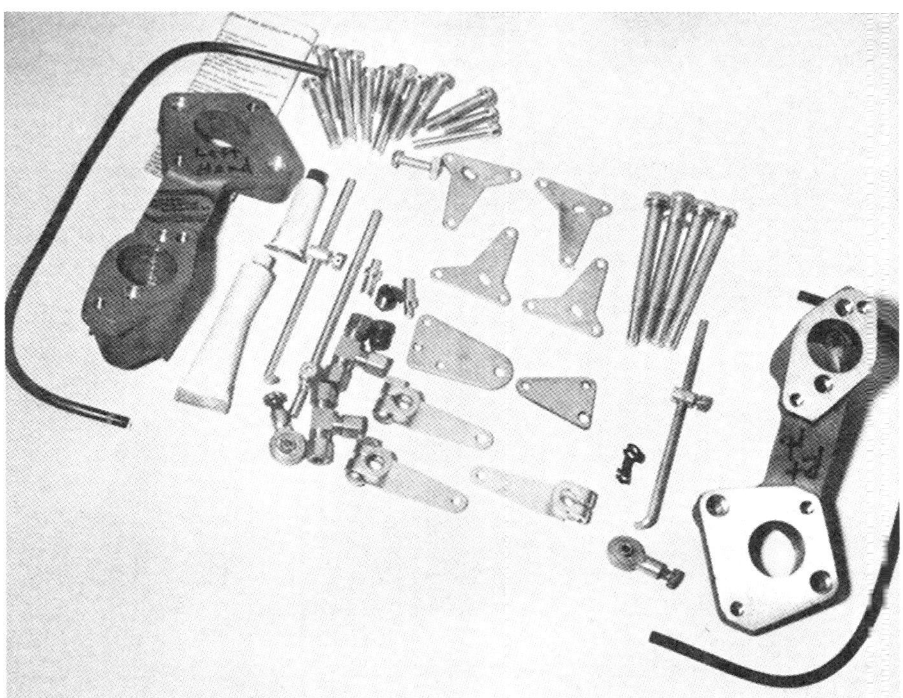

Ansen 4 × 1 conversion kit will add at least 15 HP to any two-carb Corvair; with simple installation.

The early kit added a casting to the original Corvair head. It was bolted in place, machined through, then glued, and bolted in place.

stuck shut unless full throttle is used often. That shouldn't be a problem after you experience the fun of the throttle response! There are two different stock 140 HP linkage setups in general. The 1965 system provided a true positive progressive linkage to all four carbs. GM found a downside to that, however, in that the chokes only applied to the primary carbs. On cold start up, if the throttle was pressed down, the secondary carbs would open up with no choke to enrich them—causing an instant flat spot. The natural reaction, especially among novice drivers, was to give it more throttle—worse flat spot. The late '65 (and on) linkage was more complex, adding springs and secondary lockouts to prevent the secondary carbs from opening until the chokes had already opened on the primary carbs. Many of these later setups have been bent or trashed over the years, as pieces failed or loosened. If you are starting from scratch, it is easier to build a complete early '65 system. And if you drive smart, you won't have a problem. There are also aftermarket replacement linkage kits available, for both 2-carb and 4-carb heads. They are well engineered and reasonable in price for the quality provided.

There are kits available adding adapter pads to be used on the stock 140 heads to place the float bowls forward on all carbs. The kits require all new throttle linkage. As an alternative, the jets can be relocated within the carb.

Carburetor Preparation for 4 × 1 Installation

For best results, regardless of the year of your car, your carburetors should be modified to the latest-year specifications. Even though you may feel that four identical carburetors would be especially desirable, "it ain't necessarily so!" Buy the cheapest Rochester "H" carburetors that you can find, as long as they are not corroded. The 1960–61 models can be bought at very reasonable prices. You will be making modifications on them anyway, be sure to reread the data on Improving Stock Carburetion and details on the changes which have been made in the carburetors over the past several years. Pay special attention to the details on seating the throttle plates (butterflies). Incidentally, minor and major repair kits are available from many sources.

When you have four carburetors instead of two, air velocity through each venturi is halved at any given engine speed. Thus, there is less pressure drop across the venturi to induce fuel flow. To correct for this, resistance to fuel flow must be reduced by making the jets larger, but this does not affect economy. This change also helps to compensate for the lack of a power-enrichment circuit in many Corvair carburetors. These carbs must be operated slightly rich at medium speeds to get the correct mixture at high RPM. 0.054-inch main jets are close to perfect for most engines operated at sea level; but smaller jets are required at higher altitudes. With four carburetors it is sometimes necessary to increase the idle air bleed size to as high as 0.046 to prevent an over-rich idle mixture, especially on large engines with "wild" camshafts. A fuel/air gauge is especially helpful for tuning. The new wide-band gauges are dead-on accurate and finally coming down in price. Four carburetor engines can usually use 3 to 5 additional degrees of spark advance. Whether you are using the "all-at-once" or progressive linkage, one accelerator pump can be disabled on each bank. The preferred method for doing this is to remove the internal neoprene cup, retaining the accelerator pump return spring. With progressive linkage, the pumps should work only on the primary carburetors. Some manifold-kit makers suggest straightening the accelerator-pump-linkage rods on all four carburetors to reduce pump delivery. You may want to try both methods and determine which works best for you. The accelerator pump cups are now available in Viton, a more chemically resistant material. If you are rebuilding any H-series Rochester carburetor, using the Viton cup will provide better protection from ethanol-blend gasolines, which have been known to attack other materials.

The stock 140 HP linkage has always had issues. Unless used regularly, the secondary carburetors will stick shut, or even plug up with deposits and dirt. Many 140 HP motors today are not even actuating the secondary carburetors. The Crown linkage above was marketed as a replacement for the poor GM linkage. Because it opens all four carburetors at the same time, it requires idle circuits be present on all four carburetors. This is usually accomplished by replacing the secondary carburetors with another pair of primaries. A non-progressive linkage, like the above, can make for awkward around town driving, but nice full-throttle runs. Race cars rarely sit at part throttle for very long.

If the chokes are eliminated, as they are often not necessary on the 4 × 1 setup, plug the choke shaft holes to ensure that all air entering the air cleaners will be cleaned. Plugging these holes is also necessary to make the Uni-Syn synchronizing gauge work correctly.

Control-lever changes should be accomplished by following the installation instructions of the kit you have purchased! If you intend to modify the stock levers instead of adding special ones, then these levers should be annealed prior to drilling, tapping, sawing, or any other work. They are hardened to reduce wear. Annealing is accomplished by heating the levers until they are red hot, then allowing them to cool in air. Do not try to speed up the cooling in any way.

Get Stock-Engine Flexibility from Your 4 × 1

Why add carburetors at all if increases over stock carburetion cause problems? Because stock Corvair engines are under-carbureted and cannot provide the best mid-range and top-speed performance. Careful venturi sizing ensures that air velocity through the two stock carburetors will always be sufficient for adequate fuel/air mixing-even at the lowest engine speeds. You should be familiar with the net results: smooth idling, terrific low speed punch, a decent (but not outstanding) mid-range and very little top end performance. This carefully designed setup reduces the top-RPM horsepower because there is an excessive pressure drop across the venturis—that is, the pressure within the manifold should approach atmospheric pressure at full throttle under full load. This is measured in inches of mercury (Hg) with a vacuum gauge connected to the balance pipe connecting the two manifold chambers. According to experts who have researched the problem for years, ideal racing carburetion should show a pressure drop not exceeding 1.0 inch Hg at full throttle under full load. But racing and transportation cars are two totally different breeds of cat! The racing engine typically exhibits rough, lumpy idling. And, it gives you miserable low-end torque coupled with poor engine flexibility—which means constant gear shifting to keep the engine in the usable speed range. Road cars with such characteristics have to be "rowed along with the gear lever" in traffic. "Usable" means speeds above those RPM at which the air velocity is sufficient to ensure good fuel/air mixing and consequent good distribution. If you have ever owned a high-performance sports car, you will immediately recognize these symptoms. Bill Fisher's Super 90 Porsche (in its pre-Corvair-engine form) had engine inflexibility which required keeping the engine speed above 3,000 RPM at all times for decent performance.

Detroit engineers are excellent compromisers. They have found that fine low-end punch with an excellent mid-range and better-than-average top speed is provided by a carburetor system giving a minimum pressure drop not greater than 3.0 inches of mercury at wide-open throttle. Racing carburetion systems are usually set up to give 1.0 inch Hg or less at WOT. But, when you add carburetion capacity to reduce the drop towards the 1.0 inch figure, you add more problems than performance unless you can make the additional capacity function in variable fashion. This is because adding carburetors or increasing venturi size reduces air velocity through the carburetors. When velocity falls below a "critical" value at which fuel and air cease to mix in the proper proportions, the engine is over-carbureted and produces little power at or near that RPM where the critical value was reached. Undesirable results include: rough idle, poor low-speed torque, weak mid range, and a powerful top end once RPM is increased to the point where fuel and air are again properly mixed. Detroit first solved their problem with the large four-barrel which provided carburetion equal to engine needs at all speeds: two venturis for idling through mid-range RPM, and two additional venturis which come in for top RPM. Electronic fuel injection, standard on all cars today, was the ultimate solution.

Variable carburetion combines the under-carbureted stock-engine low-end "punch" with the mid- and top-end performance of the many-carburetored setups. Hook up your four-carb linkage progressively so that the engine idles on all four carburetors, runs on two carbs until they reach from ½ to ⅔ open, and then brings in the secondary carbs so that all serve the mid and top ranges. You get 90 percent of the performance that you could expect from six carbs—or even fuel injection—and engine flexibility is retained.

A new-design, aluminum, progressive-linkage kit is now available from Roger Parent for both the 1965 and 1966–69 four-carburetor motors. It is all new and uses spherical bearings for adjustment. He also supplies replacement bell-crank assemblies for the transmission-mounted throttle rods.

Multi-Barrel, Center-Carburetor Manifolds

A single two- or four-barrel carburetor mounted on a manifold at the center of your Corvair engine can provide particular advantages, especially at highway speeds. After fighting the linkage problems and synchronization headaches of the two- and four-carb Corvair engine installations, Bill Fisher said he would never consider changing back to those complexities again. His engine had the IECO centered four-barrel manifold with four "runners" to mate with the

Two- and four-barrel carburetor manifolds are available for the Corvair. Unit here is popular IECO unit mounting Rochester 2-GC with 1.88 square inches of venturi area.

four carburetor mounting pads, This type of manifold can be recommended for use with the small-valve heads if you will add an extra carb-mount pad on each cylinder head's manifold. If you want to stick with the two stock openings, then consider using a centered two-barrel carburetor instead. Additionally, the four-barrel has progressive linkage built-in. Primary venturis provide for most driving situations, and the secondaries open at about 2/3 throttle when you really need them.

Bill Fisher's own engine was equipped with a Holley 450 CFM carburetor modified per Doug Roe's suggestions. This is the small venturied List #3367 that has 1-½-inch throttle bores: 1-¼-inch and 1-5⁄16-inch venturis. The reason for recommending the Holley over the Carter AFB is that the Holley is much better in its freedom from turn cut-out problems. This can be extremely important if you plan any slalom or hill-climb racing. Also, it can be equipped with center-pivot floats by merely installing the Corvette float bowls and floats. Holley has kits for the "4150" series carbs. In the many years since four barrels first were installed on Corvairs, Holley has released hundreds of different small four barrels. There are several different models of the 390 CFM carb available. Holleys are identified by the "list" number stamped on the air horn on most models. List number 6299 is a commonly available 390 CFM Holley. List numbers 0-8007 and 0-80507-1 are both current 390 CFM models. The List 0-6299 was an older 390 CFM carb. All have been successfully installed on Corvairs. There are no easy bolt-on center mount solutions. Tuning and modifications are necessary to complete the installation. There are several books available on setting up Holley carbs.

Centered four-barrel and two-barrel carburetor installations usually experience idle "tip-in" problems when these large-engine carburetors are used on the small Corvair engine. Simple modifications suggested by Doug Roe, the noted Arizona Corvair racer and tuner, can eliminate this problem and add a great deal of pleasure to the use of these big carbs. He suggests lowering the float level, opening up the holes which feed the idle-air needles until the needles will just seat, and increasing the taper on the idle-air needles. This increases the idle "signal" and the increased taper permits finer adjustment. The vacuum-operated enrichment (power) system should be rendered inoperative. (This is the "Power Valve" on a Holley. Summit Racing sells a plug.) When the vacuum-operated metering rods on the Carter AFB seem inactivated, tune for correct mixture with different rod diameters. The Holley carbs dominate the center mount installations because of their flexibility and parts availability.

It should be noted that center-mounted carbs suffer from a lack of carburetor heat in cold climates, and this can be a problem. And, despite the hype, the four-runner manifolds are not ram tuned.

This center-mounted Holley 4-barrel is topped with an "extra-flow" air filter. No restriction here!

Clean Carter AFB center-mount installation. Smallest available version is 500 CFM.

Streetable version of LeVair one-into-six kit. Heat can be added to the area below the carb to minimize carburetor icing in humid climates, not shown.

Dave Clemens's race car uses a full-up, racing-prepped Holley on the one-into-six. Note absence of alternator and presence of roll cage.

Another center-mount system is also available. LeVair Performance products sells a six-runner center mount system which includes head modifications to provide a center plenum mount under the carburetor and feed each cylinder with an individual runner. Several hoses and clamps are used to provide sealing to the heads, where aluminum tubes are welded on and ported for each cylinder. The system works very well, when installed and tuned correctly. Warren LeVeque recommends a modified Rochester Quadrajet 4-barrel with this setup. There are new Quadrajet replacement carburetors now available in the aftermarket.

Another option for the center-mount application is the new carburetor replacement throttle body kits from Fitech and Holley. Available as both 4-barrel and 2-barrel replacements, these use a center-mounted throttle-body containing most of the needed sensors, as well as the Electronic Control Unit. The only remote inputs are an O2 sensor, an RPM sensor, and an engine temp sensor. That is usually a water temperature sensor, but the input can be fooled by correctly mounting a sensor on the cylinder head. The fuel atomization from the fuel injectors should be more effective than the carburetor, which should help keep the fuel as a vapor on its way to the chamber.

Performance Air Filters

Corvairs came with two different air filter designs, three if you include the unique design for the Turbo. Although some early models used clean and re-oil filters as original, most of these have been replaced by the more common pleated-paper filters. The filter used on the late Corvair (most 1964-up) is common to many Chevy V8 trucks of the era—it is plenty large enough to feed the Corvair motor. There are several brands of replacement filters that are designed to clamp onto a round flange carburetor inlet. A pair of these, or two pairs for a 140, can easily be adapted to provide a "racier" appearance and allow easier access to linkage and other Corvair engine pieces. The stock Corvair carburetors, with the exception of the

Modifying the tops of the carb for a round, clamp-on filter. The carburetor will allow any clamp-on filter designed for a 2.25" round flange. K&N filter assembly Part Number RU-0620 will clamp onto the top of the modified carburetor. After this modification, the '64-up air filter housings will still attach with the stock J-hooks, the early-style snap-over clamps will no longer work.

Individual filters installed on modified carburetors.

Use a hacksaw to roughly trim the tabs.

Smooth the rough cuts to match round.

Completed preparation for clamp attachment.

1960 models, have a slight bulge on the sides of the carburetor tops. With a slight bit of hacksaw work and a smoothing file, the carburetor top can be made round and ready for an easy install. If you don't want to slice on the carburetor top, you can try slicing the clamping flange of the filter. Other aftermarket air filters and housings have been available with larger flanges and bigger replaceable elements. Some of these present a flat flange around the carburetor top which may be used to route the PCV system feed into a place similar in design to stock. You may be tempted to just clamp a small filter element onto the vertical tube from the PCV system. If you are still running a stock "fresh-air" heater or defroster, do not do this. You do not want these fumes being drawn inside the car as part of the heater feed. You can shorten the tube and route the emissions into the driver's side filter housing. This maintains the EPA legality, as any blowby would be recirculated back into the motor. As an alternative, you can route a hose outside of the engine compartment, the classic vent tube. Hooking it to the air filter housing provides clean air to the system when it needs it and a path back into the engine when fumes come out.

Weber 3-Barrel Installations

Another "carburetor" modification receiving more interest than ever, is the conversion to Weber 3-barrel carburetors. I cover the needed cylinder head modifications in that chapter, but I will speak about the conversion. The Weber carburetors came in two different sizes and are named by their throttle plate sizes. The 40 IDA and 46 IDA identify the different sizes. The only real application for the 46 IDA Webers would be an out-and-out race car. All Webers are used as pairs and there is a left and a right carb, based on the accelerator linkage. The carbs are stamped "40IDA3C" and "40IDA3C1." There are also 40 IDT and 40 IDS Webers available. They are functionally the same as the IDA. In addition, the Porsche 911 models, besides donating the 40 IDA, also provided pairs of Solex 3-barrels and Zenith 3-barrels. The mounting on the manifold face is the same on some of these, but the carbs are quite different, and are neither as popular nor as tuner friendly as the Webers. The Weber 3-barrels have replaceable venturis (Weber Guys call these "chokes") so you really can tune your carbs to meet your unique circumstances. Because each barrel has about six things that can be changed and you will need six of each item to make the

Nice Weber 3-barrel installation. Uses pressure regulator to split the flow to the two carbs. Note the updated alternator and Corvair Underground fan belt idler.

Weber carburetors and manifold. Note how the manifold bolts onto machined head.

change, any trial and error changes are expensive. All indications are, however, that the carbs run okay right out of the box, if the applicable Porsche 911 jetting is left intact. There are Corvair people who have made the investment in time and money to zero in on the right jetting, although engine modifications, such as cam and exhaust changes will affect the settings. They will usually be happy to share their suggestions. If you really get lost, try to find a Porsche mechanic who will work a "tune" on your engine, you will have the best luck with the Porsche racers. A shop that specializes in the Webers may swap out parts until it works right, then charge you just for those final Weber parts.

Although the Corvair engine is larger than the "donor" Porsche, the Porsche had a higher operating range than the hydraulic-lifter equipped Corvair, so the actual airflow tended to equal out. There are several sources for air filter assemblies for the Weber carbs. As installed in the Corvair engine compartment, you have plenty of height for taller filters. Linkage is now available from several sources as well.

For complete details on ram tuning intake systems, refer to the book by Philip H. Smith and Dr. John C. Morrison, *The Scientific Design of Exhaust & Intake Systems*. Perhaps one of the best single articles written on the subject was by Roger Huntington in the July 1960 *Hotrod Magazine*, "That Crazy Manifold." July and August 1964 *Hotrod Magazine* issues had two further articles written by Dr. Gordon H. Blair, PhD. All are worth reading.

Fuel System

The Corvair fuel tank is perfectly located for crash protection and occupant safety. From 1961 on all cars used the same tank design and new replacements are still available. Cars that have been sitting for a long time can suffer rust-out of the tank bottom. Although repair is possible, for maximum life a replacement tank, perhaps with additional rust protection, is recommended. A suspect tank can be cleaned and resealed at most radiator shops. Inside the tank, the fuel inlet and fuel level sending unit assembly are still available. The floats can leak—one symptom being a full tank and an empty indicator. Corvair fuel pumps used to be a non-issue. The basic design, a pushrod driven off a crank eccentric was simple and reliable. Occasionally a check valve, integral to the operation, would stick or pop out. The car would run out of gas and stop or flood, depending on the failure. At some point since the last Corvair was built, one or more of the companies building fuel pumps for the Corvair substituted a different material for the twin swatches of sealing gaskets used inside the pumps. The replacement material was found to lack reinforcement needed to perform

Another of the special carburetion systems utilized by GM in their experiments on the Monza S.S. roadster show car.

The carburetors look identical to Webers, but are GM engineering castings mounting Weber components.

adequately for very long. The rupture of either gasket would cause a failure of the pump. One failure mode would allow fuel to exit the lower portion of the fuel pump and enter the oil crankcase of the motor. The gasoline would both dilute the oil, and cause an over-full condition. Engines do not produce motor oil. If you ever "dipstick" your motor and find more "oil" than before, do a sniff test. If there is any whiff of gas in the oil, chances are the pump has failed and it is time for both a pump replacement and an immediate oil/filter change. The bad pumps long ago cleared out of the parts replacement pipeline, new ones from the Corvair vendors should perform well. Pumps that are left over and sitting on a friend's shelf for many years may or may not be okay. Before you install a new pump, be sure to "snug-up" the screws that clamp the pump into an assembly.

Fuel Lines

The fuel tank in the front and the fuel pump in the back might seem like a

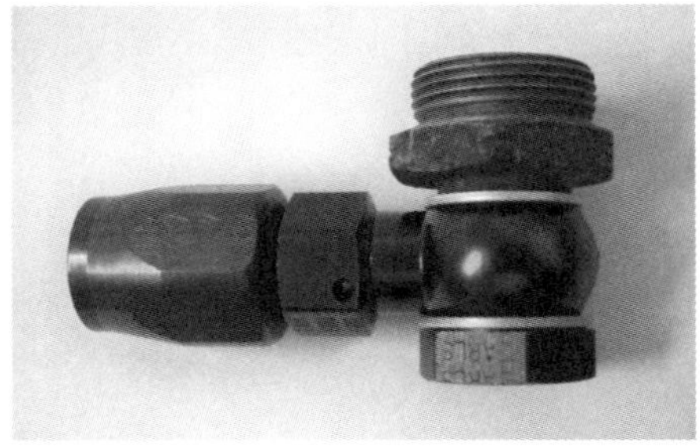

This banjo-style fuel inlet is adapted to the Corvair inlet fitting. This style inlet fitting allows clearance for the additional carburetor on the 140. Direct fit hardware is also available from Earl's Supply.

design to create problems. But most regular cars of that era had the tank in the back and the fuel pump in the front. Both implementations share a problem. The lines from the tank to the pump are under no fuel pressure during operation, actually slightly less than atmospheric with the fuel pump sucking on the end. The Corvair has two places where fuel hose connects hard lines in the sucking state. Just behind the fuel tank and just over the rear wheels where the fuel line into the engine compartment mates to the body line. Any cracks in the hoses or loose connections can allow air into the system and a low fuel condition in the carbs. Check the condition of the hoses, especially on a car that has been sitting for a while. Replacement is good insurance. Screw clamps are suggested. A stock Corvair fuel line is large enough to feed several hundred horsepower. Larger lines buy nothing.

If you want to replace the fuel lines in the engine compartment with something with a snazzier appearance, you can. The Rochester H carb has the same carburetor inlet thread as the standard center-pivot-float Holley. Earl's Supply and other sources have the parts to provide all the braided line and anodized fittings that your budget can afford. During the design phase, remember that the two carbs on each side sit pretty close to one another. Allow space for the fittings and hoses.

I have run both Holley and Carter electric fuel pumps on race cars. I finally standardized on the Carter rotary pumps (P/N 4594) as a reliable workhorse. It has never failed to work. If you are not very careful about where you mount electric pumps, you will get a droning sound in the cockpit, acceptable on a race car, not so on a street car. The best street installation I have seen was the mounting of a low-pressure pump inside the tank. With the fuel dampening the sound, it was quiet and would have appeared stock. In the Corvair world, there is an ongoing conflict on whether to use the stock mechanical pump in the stock location, or replace it with an electric pump mounted near the front of the car, near the tank. There is plenty of ammunition on both sides. The stock pumps, by now, have many millions of miles of use, and a proven design. There are known failure modes and sufficient troubleshooting procedures are available to assure proper operation or correction. Electric pumps have come a long way in 40 years, and, with the proper precautions, can perform well in the Corvair. Noise can be an issue, a solid mounted pump on the floor will sound like someone hammering on a drum under your feet. Isolating the electric pump, wiring it in the proper/safe way and repurposing the fuel line system to be a pressure-based system instead of a vacuum-based system, by verifying hose and clamping condition, will result in a safe, reliable system. Do not mount an electric pump in the engine compartment. All of the manufacturers recommend a low mounting near the tank. One extra benefit of the electric pump design is better restarts after sitting for a long time. Even if all the fuel evaporates from the carbs, 20 seconds of electric pump operation will refill them for you, and the engine will start right up. It is true that electric pumps can fail, too.

Electronic Fuel Injection

Although the Corvair engine, as supplied by Chevrolet, was never equipped with fuel injection, Chevrolet's love affair with fuel injection actually predated the production of the Corvair. Chevrolet delivered some production Corvettes and full-size Chevy cars with fuel injection from 1957 through 1965. The R&D labs at Chevy installed at least one of the FI systems—a Rochester Division product—onto a Corvair motor. Like using the turbocharger that eventually made

Clark's newest fuel injection kit is complete. A high-pressure fuel pump and return line are purchased separately.

GM throttle body is adapted to Corvair mounting pad.

The kit uses reliable GM wiring and controls. Along with the fuel injection system, the kit replaces distributor with a GM coil pack ignition system. The crank pick-up is mounted on the original fuel pump locking bolt. Electric fan is not a kit item.

This is an original Rochester Fuel Injection system installed by the factory for racing and testing.

it into production, Chevy was looking for ways to make the Corvair engine more competitive in the horsepower department. Production use of FI was widespread by the 1970s, especially on the smaller performance car engines. Datsun changed the 280Z over to FI in 1975, Porsche had already changed the 911 series. Shortly after that, enterprising souls were adapting those Bosch FI units to the Corvair motor. The Corvair's integral intake manifolds and the air-cooled nature took some creative adapting for systems originally intended for water-cooled engines, like the Datsun Z. Technical literature at the time and high prices for the various FI piece parts also slowed development. Of course, we all knew carburetors pretty well, and the new fuel injection was now electronic—EFI. The "electronics" part gave many of us shade-tree mechanics major concerns. But the automotive industry had really completed the changeover to EFI by the late '80s. Test procedures, wiring, and hardware availability became easy to find. Importantly, the aftermarket hot rod market was beginning to shift to EFI, offering easier to install systems. The commonality of "conversion" to EFI for street cars made Corvair conversions more popular. We are still adapting systems and pieces from other cars—that is the inexpensive way to go—but the electronic expertise and knowledge is much easier to find today.

For racing, Corvairs have been slow to move to EFI because of the rules for most of the classes in which they compete. Vintage racing means vintage, in most cases, forget EFI. The Corvairs in road racing have always been regarded warily by the rules makers, who have always been frugal in the changes they have doled out to the Corvair.

In autocross/slalom competition, SCCA has finally allowed EFI in the Prepared Classes, without the former weight penalty. This is not a gift to the old cars, it is an attempt to get newer cars—all of which are EFI equipped—into the classes. All of the racing EFI systems require machine work on the Corvair head to adequately locate the "injector-per-cylinder" needed for performance use. This is the big differentiator between racing and street systems. The racers aren't too concerned about cold starts or fuel mileage, two things that endear EFI to street drivers.

All fuel injection systems today have two main parts, the air/fuel delivery hardware and the control/sensor package which measures the air and measures out the fuel. There are at least a dozen aftermarket control systems available which can read engine performance and optimize a system for competition. The sensors for the Corvair are the same as for other engines. Locating them on the Corvair is a tricky bit, because air-cooled motors operate so differently than water-cooled motors. The factory units from GM and Ford are very sophisticated and, only recently, have been "cracked" to provide control of the sensor input. The modification of the factory computers is still somewhat of a cottage industry, with users trading notes and tips. The built-from-scratch aftermarket units are very well documented and allow you many ways to adjust your systems for changing inputs. The newest of the systems now self-configure during installation, bringing us closer to the plug and play EFI system. For air and fuel delivery, the part of the system which will be unique to our Corvairs, a popular system is to follow in the Porsche footsteps and use EFI hardware which was designed to replace the Weber 3-barrels on the Porsche—those carbs that we were already adapting to our Corvair motors! The BORLA throttle bodies will drop on in place of the Weber 40IDA 3-barrels and provide plenty of airflow to a ported Corvair motor. Throttle linkage and air cleaner assemblies are the same as used on the Webers. All fuel injection systems run at much higher fuel line pressures than carbureted systems. Most new cars place the EFI fuel pump inside the fuel tank. If you decide to do this, be sure to equip your installation with a good high pressure line and hose system, both for the pressure feed and the return line, if used. When you are serious about the use of EFI on your race car, check with the manufacturer of the control system to see if they have an authorized installer near you. This could save you a lot of work with the final tune. You might be able to bring them a partially completed install with their expertise used for the final connections and settings.

Street EFI systems are becoming more common. Clark's Corvair formerly sold a very complex system which, while it worked well when installed correctly, took a lot of patience and planning to complete. Recently Clark's has been selling a new street system, a throttle body system requiring no cylinder head machining. The first systems are replacements for the two-carburetor factory engines, up to 110 HP. This system is assembled from all GM parts, including the Electronic Control Unit, which has been programmed for the special needs of the Corvair motor. The system retains the look and feel of the Corvair motor, adds the cold start and gas mileage benefits of modern fuel injection and tops it off with GM reliability. These systems are on the road and working.

BORLA throttle bodies replace the Weber 3-barrel carbs and use the same air filters and linkage. The control system runs both the injection system and the integrated ignition system.

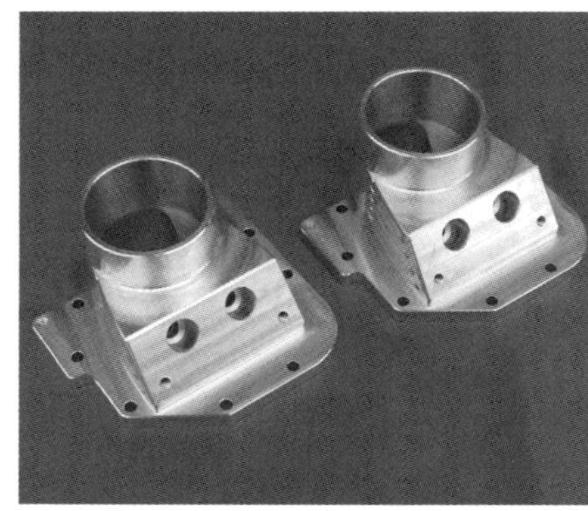

These injector housings replace the top of the Rochester H/HV carb for an EFI conversion. One or two injectors can be used, depending on the electronic controls.

Black Hawk Engineering is developing a bolt-on EFI system specially designed for turbocharged Corvairs. The injectors bolt onto the turbo inlet in place of the Carter YH carburetor, and use the Corvair air filter assembly. It still needs data inputs to control the injectors, but the growing availability of EFI components for almost any engine makes this an inevitable conversion.

Ignition

Your Corvair ignition system will likely be adequate, even for serious racing applications. Simple modifications can adapt it to complement the changes that you make to your engine. Corvair distributors vary widely in their advance-curve characteristics, but the 1965–66 manual transmission 140 HP unit, Part No. 1110330 is almost perfect for peak performance when used with a non-supercharged engine. Unless electronic controls are added with a supercharged or turbocharged motor, the original pressure retard equipped distributor should be maintained. Certain early distributors should be replaced because design faults make them unreliable. Late smog pump compliant distributors have advance curves that were bizarre at best, and can be virtually unusable on 35-year-old motors. Sparkplug selection is important for performance, and types for high-performance, high-compression engines are discussed. Two bugaboos of high performance are detonation and preignition. Both of these are discussed in this chapter.

Glamorous transistor ignition systems are interesting technical achievements, but they can also be expensive. Their advantages of long point life, easier starting in cold or damp weather, and possible increased sparkplug life are real. But, by themselves, they do not increase acceleration, add top speed, or give more gasoline mileage when compared with a stock system that is operating as it should. The peak ignition requirements are for maximum voltage at the plugs when the engine is being accelerated at low speeds under full throttle, according to Champion engineers. Transistor systems are equal to, but not better than, the stock system under these conditions. The best attribute of the aftermarket systems is low maintenance. In theory, you should be able to install one of the units and support it like you would a new car—leave it alone for 50 thousand miles.

Distributor in foreground with screw-attached cap was introduced in 1962, to make all six-cylinder Chevy ignitions use similar parts. Earlier model lacked bushings in housing, had clip-attached cap. Cone-shaped device on distributor is vacuum advance (retard mechanism on Spyders).

Material presented here does not cover the entire picture, so you should also read the Engine Electrical section of the Corvair Shop Manual. It contains many important details.

SPARKPLUG INTERCHANGE TABLE

Application	AC Original	AC Optional	AC Current	Champion	NGK	Bosch	Denso	Motorcraft	Autolite
Low Performance 80-84-95 HP	46FF		46F	L10	B4HS	W9AC	W14F-U	AE6	275
Alternate (Resistor)		R44FF			BR4HS		W14 FRU		
High Performance 98-102-110 HP	44FF		44F	RL87YC	B5HS		W16FS-U	AE4X	275
Alternate (Resistor)		R44FF			BR5HS		W16FSR or 13033		
Special High Performance 140 HP		R44FF		L87Y	B5HS	W8AC	W16FS-U	AE4X	2656
Alternate (Resistor)			R44F		BR5HS		W16FSR or IWF16		
Turbocharged 150-180 HP	44FF		44F	L87Y	B6HS	W7AC	W16FS-U	AE3	2656
Alternate (Competition)		R42FF		L5			W20FS-U or IWF20		

Note: Several manufacturers have many additional options.

The stock ignition system consists of a distributor, coil, sparkplug cables terminating in access-cover boots to seal the blower shroud from cooling-air loss, and sparkplugs. The Corvair distributor has a single point set and condenser attached to a movable point plate which is advanced or retarded by a vacuum-advance diaphragm connected to intake-manifold vacuum. This unit mounts on the outside of the distributor. A similar-looking device provides "pressure retard" on the Spyder models. A rubber boot covers the mounting arm to prevent water from entering the distributor. Centrifugal (mechanical) advance is provided in the distributor by weights that advance the distributor cam as engine RPM is increased. The coil is mounted at the rear of the engine alongside the distributor, except on smog pump and air-conditioning equipped models. Two basic distributor designs have been used, with the '60–'61 models considered unreliable because their shafts run directly in the aluminum housings and are prone to seize. The later, 1962–69, design is well proven, easy to modify, and has many parts common to other GM 6-cylinder engines of the same era. It uses a single bushing in the revised aluminum housing, relying on the locating hole in the rear engine housing to retain the other end of the driveshaft. Unlike many other cars, the Corvair distributor is gear driven directly off the crankshaft. This stabilizes the spark output because it avoids timing chain slop common to most V8s. There are aftermarket suppliers who rebuild the distributor housings, adding either a second bushing or a sealed ball bearing (my choice) to further stabilize the spark. The other advantage of the sealed ball bearing it that it keeps oil from building up inside the distributor. This can be a problem in high output motors with limited ventilation capabilities. The point plate can also be reworked to provide a more solid rotational base for the points. The stock Corvair plate just used a pin stuck through a punched hole. In the 1970s, Chevy produced an updated point plate for the early (1971–73) Vega motor. These worked fine on the Corvair, but GM discontinued that plate many years ago.

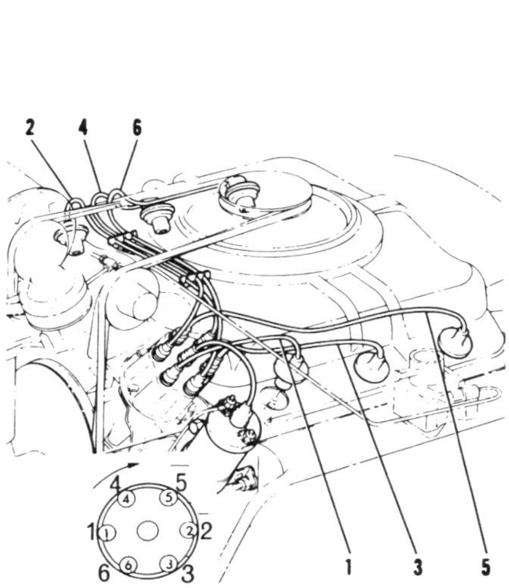

Corvair firing order is 1-4-5-2-3-6.

High performance breaker points are available to fit the '62–'69 Corvair distributor, due to the commonality with the other GM 6-cylinders. Accel, Standard Motor Products (in their "Blue Streak" line), and others have produced better quality points, rotors, and caps. A note of caution here. If you install one of the high RPM point sets, all of which use a heavier duty spring, it will cause an increase in side load on the mounting location on the points plate, as well as a small side load increase on the distributor shaft. The original style points plate will suffer a dramatic decrease in service life. If you have updated the distributor to a double-bushing or ball bearing and added the bearing type points plate mount, you should have no problem. If you have the distributor apart for rebuild, you may want to add a shim or two to take up some of the vertical play in the shaft inside the housing. The shims will go just above the upper roll pin. Some play is needed, though. It shouldn't be hard to turn the shaft.

Hopped-up Corvair engines can perform quite well with the stock ignition system. Although numerous stock mechanical and vacuum-advance curves were used, each tailored for a particular transmission, displacement, compression ratio, and camshaft, all allowed for sloppy driving and maintenance habits. The factory engineers assume that you will permit the engine to lug, will use poor gasoline, and will probably let carbon deposits build up.

Greatly increased advance in the lower-RPM range can usually be used in cars that will be primarily engaged in acceleration contests (drags) or autocross. This is because the engine will pass through the low RPM range very quickly, or will seldom be operated there at all. Also, long-overlap cams pump part of the fuel charge back into the manifolds at low speeds, even at full throttle; reducing charge density and allowing more advance to be used. Thus, mechanical advance-curve changes can provide startling full-throttle acceleration improvement. A total centrifugal advance of 16–18 degrees should be "all in" by 3,000 RPM for best performance in most engines. Springs to permit this advance can be Part No. 1965545 from the 140 HP distributor, or you can experiment with a single spring, or stretched stock springs. Several aftermarket manufacturers sell spring "kits," with several pairs of color-coded springs to try. Reduction of the advance to restrict the total to 16–18 degrees engine (8–9 degrees distributor) is illustrated in the photo. By adding the sleeve over the pin, you change the starting point of the mechanical advance weights, reducing the total amount of mechanical advance in the distributor. When you install the modified distributor, the initial setting can be increased by the number of degrees the total advance was reduced—at least for a trial setting. Example: 28 degrees original engine (crank degrees) advance reduced to 18 degrees is a 10 degree reduction. If original setting was 6 degrees BTDC (crank), add 10 degrees to 6 degrees for an initial trial at 16 degrees BTDC. Experiment with the initial setting to determine the optimum position just short of detonation. Be sure to make your checks with the muffler installed so you can listen to the engine. Retard the spark several degrees when changing to a free-flowing or open exhaust system because the better breathing exhaust may lean out the mixture in the cylinder. What you are trying to do is to get maximum power by establishing the ignition-advance curve to keep the engine just short of detonation right through the RPM range when operating at full throttle. If you want to keep good gas mileage as part of the performance equation, you will need to limit your initial advance because at part throttle, light load, the vacuum advance becomes additive to the initial and any mechanical advance already there. Stock late model

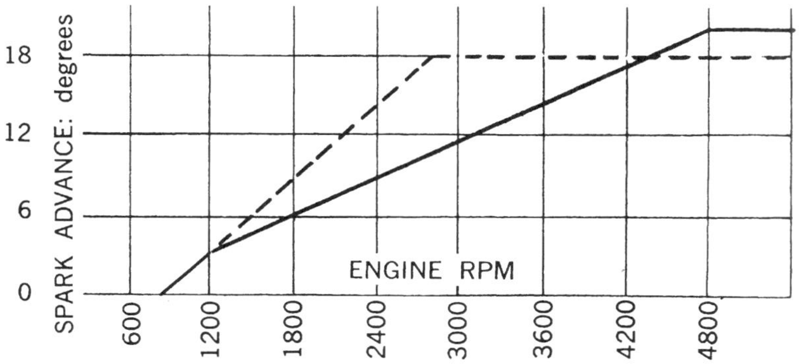

Centrifugal-advance curves for 1965 110 and 140 HP engines. Dotted line for 140 HP model shows that it is nearly perfect for acceleration as described in the text.

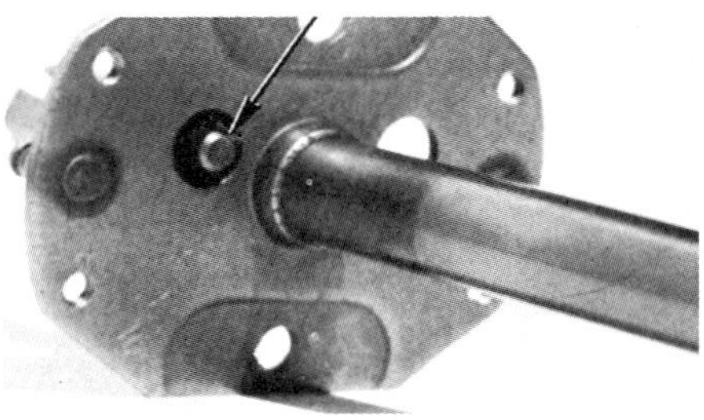

Arrow indicates cam-stop pin which can be metal-sleeved or brazed to larger diameter to reduce total centrifugal advance.

110/PG cars are prone to detonation under these circumstances. (But can get great gas mileage!) Most Corvair motors in competition can use a total of 34 to 36 degrees of timing advance. This is the total number of degrees at the crank, including initial advance and mechanical advance. Vacuum advance is largely irrelevant under full throttle, the low vacuum makes it disappear. If the distributor is modified to provide 18 degrees of mechanical advance all in by 2500 RPM, similar to the performance of the stock 140 distributor, the initial advance should be set at 16 to 18 degrees. Turbocharged motors and some smog motors had mechanical advance indicator plates on the rear housings, which showed up to 24 degrees of initial advance.

Coil output, at least on non-Spyder models, can be improved 20% by merely installing the 1962–63 Spyder coil. However this coil, marked 091-12V, is no longer supplied by Chevrolet. It was also used on some Corvettes. There are many aftermarket high winding-ratio coils available. Be sure the coil is compatible with your trigger mechanism, whether point-type or otherwise, and the voltage you are feeding it.

Stock-heat-range sparkplugs will have to be replaced with colder ones for racing only usage. Cold plugs should not be used for normal street/highway driving as the engine will spit back excessively, plug mileage will be drastically reduced, and gasoline mileage will suffer. The original AC Delco plugs for the Corvair, the AC 44FF and 46FF plugs and their R (resistance) equivalents are only available via the swap meets. There was also a 42FF plug. It was recommended for competition use on Turbocharged Corvairs. But there are many workable solutions available. A plug that is too hot will last only a few miles and may cause destructive preignition. A near-perfect plug (unless you have milled your heads to get the best deck clearance as described in the "Cylinder Heads" chapter) specifically engineered for the Corvair is the Champion UL-15Y. They are even harder to find than the AC Delcos. It has the projected-core nose that acts as a "warm" plug at low speed and cools off to act as a "cold" plug as RPMs are increased. It also has an auxiliary gap that increases plug life between replacement or regapping because it changes the spark-voltage characteristic. When the spark has jumped the auxiliary gap it has sufficient voltage so that the electrode path will be used, rather than leaking away ineffectually across fouling deposits which may have built up on the porcelain and shell. The Champion L-12Y was a colder plug designed for Spyders. It has the projected-core nose without the auxiliary gap. Champion now recommends their RL87YC copper core plug. There are many available sparkplugs available for the Corvair. Bosch, NGK, Denso, Motorcraft, and Autolite all have Corvair-specific sparkplugs. The Denso Iridium IWF16 is a newer plug that has worked well on the Corvair. It is an air-cooled VW application. Some aftermarket electronic ignition systems are capable of firing cold or even semi-fouled plugs. Both high-energy and capacitive-discharge systems have reserve capacities for this purpose. One other sparkplug note, the original Corvair sparkplugs were threaded all the way to the end on the plug body. All replacements, including the ACs ending in "F" instead of "FF" are a thread short. After running for a while with the thread-short plugs, the companion thread in the cylinder head will fill up with carbon. If you now stay with thread-short plugs, you will be fine. But if you score a set of AC44FF plugs at a swap meet and want to install them, you MUST use a thread chasing tool to clean out the carbon from the bottom of the threads in the head. If you do not, you run the risk of the new plug screwing tightly in against the carbon deposits then stripping out the fragile aluminum threads above them. Clean out the threads in the head first.

At the beginning of the twenty-first century, new cars have smart computers which sense the optimum time for spark and control fuel mixtures depending on what is coming out the tailpipe. The cost of adapting a full setup to the flat six Corvair motor has deterred all but the most intrepid souls. Some of the piece parts and equivalent systems have made it into the automotive aftermarket and have been adapted to our Corvair motors. The J&S Safeguard is an electronic ignition control package that can identify individual cylinder detonation and control the amount of spark advance for that cylinder to eliminate detonation. It has been tested and installed on Corvair motors, and users are enthusiastic about it. MSD Ignition components have a complete family of ignition controls which can respond to turbocharger boost and control spark advance to the entire motor.

Coming onto the market are completely digital systems that can read and respond to engine sensors and appropriately change spark and fuel mixtures. As trick as these systems are, and as necessary as their utilization may be to wring that last extra horsepower out of the Corvair motor, the stock system has proven amazingly versatile, rugged, and reliable. With a few mild updates to current heavier duty parts, a points-based system can push a Corvair motor to 6,500 RPM all day long. A last note on electronics-based, points replacement systems. The first step in installation of one of these systems should be to adjust the existing points to a correct setting and remove the points plate, with the points intact, and drop the whole system, including the thin wire to the coil, in a locking plastic bag and put it in your glove compartment, along with a screwdriver. This is your get-home kit, should the electronics system fail.

Back in the 1960s, sparkplug cables were a normal replacement item. They lasted two years in normal usage and were replaced at a tune up. Great advances in cable technology and materials have given us replacement wire sets that will last many times longer than stock. Solid core wire is never recommended for Corvair cars or trucks. The RF emissions from those cables can cause interference in any electronic component on your car, or on the car that pulls up next to you. The newest, wire-wound core, 8mm wire has only about 300 Ohms per foot and has much better insulation than original. Make sure that the sparkplug wires you use have the proper snap-in type boots. Engines to be raced should be equipped with retainer clamps to hold the boots in the shroud as these have been known to blow out when the engine is operated at high RPM. The clamps that secure the original road-draft or PVC vent tube work perfectly. Because they are no longer available new, you can use them as a guide to make a fabricated part. Install them with self-tapping sheet metal screws. Or, you can braze ¼-inch-20 nuts behind holes alongside the boot openings in the shroud and use capscrews to secure the clamps for a more finished appearance. Because the original ignition wire was so stiff, Chevy installed 90-degree plug boots at cylinders 4 and 6, turbos had one on number 5 as well to clear the turbo mount. With the latest, softer silicone cable, straight boots can be used at all positions. Keep the distributor cap and rotor and the top of the coil clean to avoid arcing. Replace burned parts. There are several aftermarket caps that use brass contacts instead of the original aluminum contacts. A good quality test for the cap is weight. The caps with the higher dielectric strength material and the brass contacts are much heavier than stock.

Do Not Dyno-Set Ignition Timing

Chassis dynamometer-set ignition timing will often be too far advanced. Such a setting causes detonation because the engine is not allowed to take advantage of combustion-chamber turbulence, and because sparkplug temperature is greatly increased by advancing the spark setting. Detonation will result in either case and can also lead to pre-ignition in the worst cases. This all happens because the chassis-dyno operator usually sets the distributor for the best flash reading, and you drive off the rollers with the engine detonating merrily. Worse yet, you may not drive the car after setting the timing on the dyno. If you enter the car in competition the chances are very good that you will not hear the engine detonating because the exhaust noise masks the detonation. The dyno operator is not likely to pay you for a blown-up engine. It is essential that you understand what can happen to your engine, especially with turbocharged models. The air-cooled engine cannot tolerate excessive spark advance as there are already enough problems involved in keeping it cool.

While it is possible to cheat a little on the factory settings (not on Turbos) by advancing a few more degrees beyond the Shop Manual recommendations, detonation must be listened for and the spark retarded if it occurs. The pre-smog control factory recommended setting is described as "spark advance for 2% power loss below borderline knock setting." That is GM speak for advance it until it knocks, then back it off a bit.

More Aftermarket Items

Mallory Electric made several models of replacement distributors for the Corvair. The earliest models featured a three-lobe cam inside the distributor, unlike the six-lobe cam used in the GM Corvair distributors. With the three-lobe design, each of the two points sets inside fired three cylinders. These should be avoided because they really need a distributor machine to properly set them up. Mallory did sell a six-lobe version for the Corvair. All of the Mallory distributors were mechanical advance only. The later units can be converted to a Mallory Unilite electronic system. Some of the Mallory units used a special two-piece cap. Replacement caps are no longer

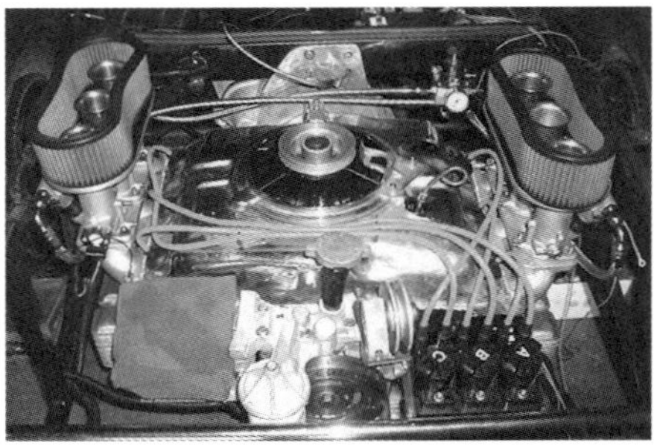

The TWM throttle bodies and the waste-spark coil packs are being driven by an SDS control unit, produced in Canada.

The plug drops into the distributor hole, retaining the shaft that is still needed to drive the oil pump.

Silicone Wire Systems 8mm ignition wires for the Corvair. These sets use correct snap-in seals at the plug.

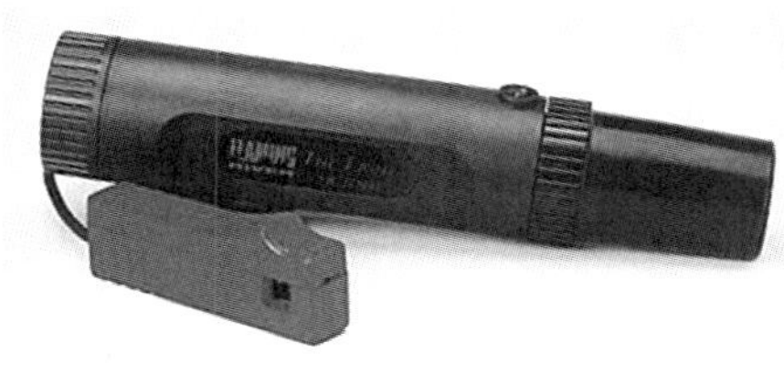

This self-contained timing light is perfect for Corvairs, especially if you have moved the battery forward to the trunk. This unit has a single clamp over the plug wire. D-cells power it from inside.

available, use with care. Most aftermarket ignition boxes will trigger with a set of points, but offer more precision and longer life if triggered via some type of electronics. There are several kits to replace the points with a magnetic pulse system, based on Chrysler components. This can be used to trigger a multitude of electronics. New items become available constantly, so stay tuned in.

The Pertronix Ignitors

The most common points replacement system is the Ignitor, by Pertronix. They have two models specific to the GM six-cylinder distributor as used in the '62–'69 Corvair, the Ignitor I and the Ignitor II, both units replace the complete points plate assembly inside the cap and hook up via two wires to the coil. Installation of either changes the initial timing setting, so initial advance must be rechecked. There are minor electronic differences inside the two units, but they are functionally the same for most of our needs. It is important that you pair the electronics with the appropriate coil. The Pertronix website has a section that covers the calculation. Also a decision point for you is whether to supply a full 12 volts to the electronic ignition module, via bypassing the stock resistance wire. With the stock wiring install, both the coil and the module are fed a resistance-dropped 8–9 volts. If you just bypass the resistance wire, you may run into problems, because the stock coil is not designed to run on 12 volts all the time. Replacement coils come in all varieties. Bypassing the resistance wire and providing a 12-volt feed must be factored into the coil calculation. The Ignitor I isn't voltage sensitive (within the 6–16 volt range) but the Ignitor II will provide a hotter spark with a full 12-volt feed and their Flamethrower II (0.6 ohms). If you are running a turbocharged motor with a stock-based distributor, this system is the easy way to go as it retains the factory pressure retard unit and its timing protection.

Because the later Corvair distributor design was used for the distributors installed on the inline Chevy six from 1963–on, the housings and other items can be adapted onto the Corvair. The inline sixes switched over to the HEI-style distributor in 1974. The MSD magnetic-pulse distributor shown in the photo has a replacement shaft installed and drives a racing version of the GM HEI module and an external coil from Davis Unified Ignition. Although the diameter of the lower housing is slightly smaller than on the Corvair, the unit will fit. The items that make the Corvair distributor unique are the main shaft and the drive gear. Those items have been adapted to an HEI distributor. On the 140 motor, the early HEI distributor's large cap will interfere with carburetors and linkage, unless installed higher up. Several companies have now introduced replacement distributors for the inline six, and those may be adaptable as well. There has been a new high-energy distributor introduced specifically for the Corvair. In "Ready-to-Run" form, it can be dropped into most Corvairs. No external box is used, just a simple 3-wire hookup (see diagram). The distributor is available with several different cap designs and colors, and, like the 1960s GM V8 design on which it was based, allows both vacuum and mechanical advance adjustments. It is also available in a direct magnetic-out version, perfect for triggering an MSD, Crane, or other external box ignition system. As of 2018, there is no pressure retard unit available for this distributor.

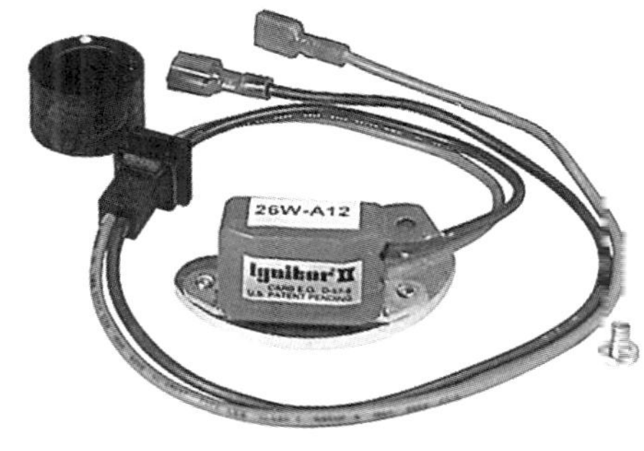

The Pertronix Ignitor II is an easy installation into the later Corvair distributor.

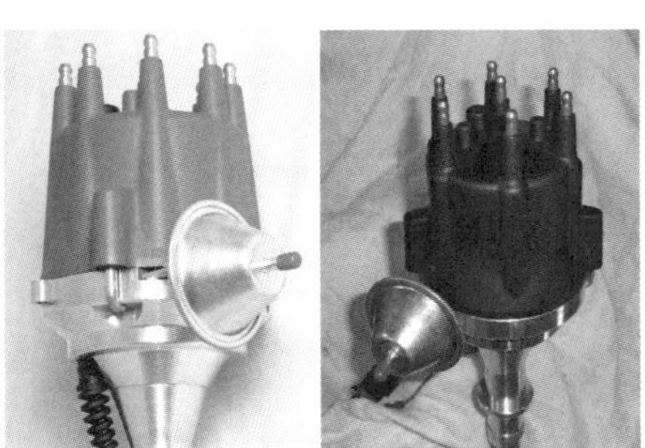

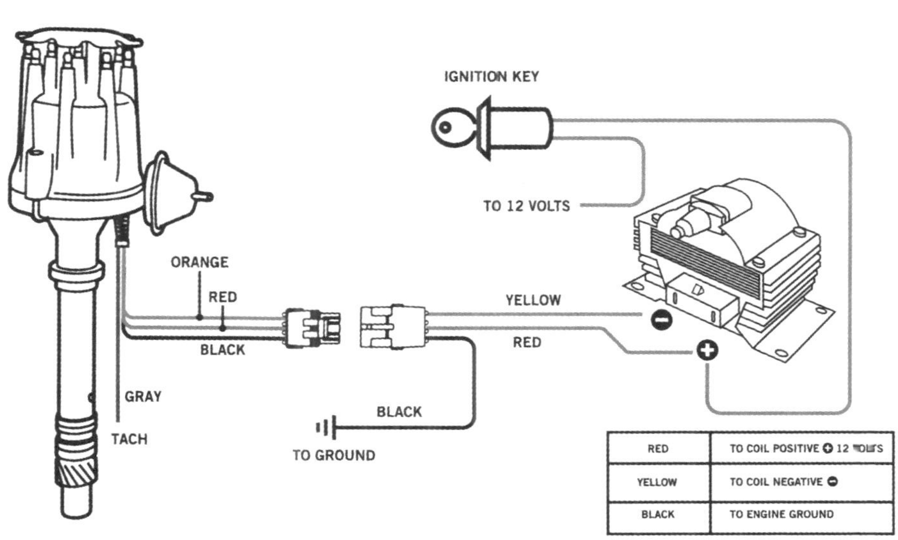

MSD distributor for Chevy inline 6, adapted for Corvair, photo and diagram. Diagram courtesy Silicone Wire Systems.

5

Exhaust

Corvair exhaust stacks are pressed straight into the head casting, with a reduction in stack cross-sectional area required to clear the pushrod tubes. This excellent design keeps ports and stacks extremely short to reduce heat transfer to the cylinder head. Exhaust stacks are connected into a cast-iron exhaust manifold that exhausts straight forward into an equalizer pipe that crosses over from the left bank, joins with the right, and continues into a muffler on the right side of the engine. The 140 HP engine has larger diameter exhaust stacks and manifolds, no crossover, but two mufflers, one at each side. The Turbo exhaust system detours through the turbo-supercharger before leaving through the larger muffler on the right side.

Stock mufflers do a marvelous job of reducing noise over a wide dynamic range of frequencies: 30 to 300 pulses/second from 600 RPM idle to 6,000 RPM. Muffling is a complex phenomenon. To truly reduce overall objectionable noise, the muffler must smooth the exhaust flow and reduce the resonance which occurs at several speeds in any system. Each of the baffles and tubes in a stock muffler has to reduce the noise at particular engine speeds. Many replacement mufflers, including all glass-packs, many large-chambered, and most "Turbo-" type mufflers nearly always resonate badly at several speeds, and often at highway cruising speed. Noise should not be lightly regarded by the owner who expects to use his car's full performance capabilities. The true enthusiast always uses as quiet an exhaust system as needed because he realizes that it attracts far less attention. A balance must be struck to reach an acceptable amount of noise inside the car because others may want to ride with you. Corvairs provide a real challenge for replacement systems because the system is really exhausting two lower-speed three-cylinder motors. The lower frequencies are hard to mask and cause droning and a "blatty" sound impression.

Power can be increased by reducing exhaust-system restrictions, or by making the restrictions work for you via scientific design. Reducing exhaust-system restrictions, at least until the '60s, meant that peak HP was only obtainable with maximum noise. Tighter laws and increased enforcement have pushed the automotive aftermarket into providing tuned, muffled systems capable of near-maximum performance at acceptable sound levels for road use. Exhausting a Corvair with such a tuned system turns out to be almost impossible because the engine location eats up the space required for one of the "basket of snakes" systems shown in the diagram. Such systems are completely described in an excellent book, *The Scientific Design of Exhaust & Intake Systems*, by Philip H. Smith. It covers the subject in great detail and is easy to understand.

Because we can't fit the truly tuned system, both for space and noise constraints, let's look at other possibilities that use mufflers, in order of increasing performance expectation.

1. Stock exhaust system with one or two stock mufflers

2. Two or four special mufflers with lengthened exhaust and tailpipes

Drawing of a tuned-exhaust system designed from details contained in Phil Smith's book. Assumes use of 4 × 1 heads with the big exhaust valves and large exhaust pipe stubs. Of interest to exhaust-system constructors: it is nearly always less expensive to buy tail pipes of the desired diameter than lengths of straight tubing, even though the tail pipe may give you more tubing than you really need.

3. Replace stock manifolds with headers, using stock or special mufflers

4. Replacement pressed-in exhaust tubes, angled for clearance, with flanges welded on

Converting any non-140 HP motor to a stock 140 HP type dual exhaust system is an effective way of relieving some of the stock backpressure, at the same time retaining the stock-type sound and appearance. Vendors offer

The metal band around the center of the donut holds the sealing material together as the manifold ring clamps down over the tube. When material is not up to the task, it blows out with the exhaust gas. I have pulled manifolds off finding only the metal rings left behind. This often occurs with headers.

systems which adapt to the slightly smaller manifold outlets. Be sure to use exhaust donuts which mate tightly to the tube. Fitting the left side muffler under the early model driver's side requires some careful parts selection and work. Because the late model was designed for it, installation is simple, and the sound is certainly no worse than the stock 140 HP exhaust. The lower HP heads can be converted to 140 exhaust tubes, but there are some intricacies in the machining.

There are several aftermarket systems produced that utilize the original manifolds and design pipes and bends to flow the exhaust as long a distance as possible before letting it out. These systems can provide a slightly louder than stock sound, depending on the muffler design used. Any performance improvement is negligible, though. The sound may be worth it for you.

Headers

True, tuned headers for the Corvair would be a logistical nightmare, with pipes going from one side of the motor to the other. Instead, the headers we have are built for ease of installation, better flow, and (sometimes) improved sealing capability. There are several other header-related issues which must be covered. The location of the Corvair header, bolted under the motor, and aiming straight down to soften the exhaust turn, is a virtual death sentence. The header pipes take hits on driveways, speed bumps and anything else the driver just clears with the front suspension crossmember. The pipe can be crushed to a "D" shape, and sometimes ripped off. Lowering the car to "improve" the handling can make this problem worse. If you want to run these on the street, be careful. Most of the headers can be installed in either a forward facing or rearward facing position. When the collectors face forward, you have room to allow a U-bend and a good sized muffler per side on the trip to the rear. Rearward facing headers are more limited, unless the intent is to make a double run heading forward with one muffler, then to the rear again. A major improvement in perceived exhaust note will result from a crossover balance pipe being inserted into the system ahead of the parallel mufflers, connecting the left and right side exhausts. This raises the frequency and changes the sound from both pipes. Getting headers (or even stock manifolds) to seal up to original exhaust stacks can be a real chore. The nice asbestos gaskets of years past are long gone from the market. Several vendors offer heavy-duty, even aluminum replacements. I think the aluminum is not up to the job, it would melt and depart in short order. But other designs, graphite and others, show some promise. If you are evaluating bolt-on header designs, I recommend using the type that presents a loose flange for each tube, rather than the welded-into-one design. With a loose flange for each tube, and the splayed tube sticking through flange, the individual flanges can adjust to the difference in cylinder to cylinder spacing, and allow the flange to "clamp" the tube onto the gasket and exhaust stack.

If noise is not a problem (as in some competition events), connect a straight pipe to the outlet of each exhaust manifold or exhaust header. For an all-out competition machine, the exhaust manifolds or headers can be installed backward, facing to the rear. Spyders run in competition can use a very short exhaust stack, exhausting straight upward through a hole in the deck lid. Very few sites allow true, open exhaust anymore, in anything other than professional competition, but this is certainly the inexpensive way to go, if you can. It will sound truly loud!

The current all-out racing exhaust technology involves replacing the cylinder head exhaust tube with a completely round, larger cross section tube, and installing it at an angle into the head to clear the pushrod tube. (Some milling of the exhaust port is required.) If done correctly, the pipe clears the pushrod tube, and doesn't restrict the exhaust flow past there. The short tube terminates in a flat, two-bolt flange, where the custom header can bolt on. An alternate method is to cut a mandrel bent U-bend in half and install the straight end into that machined port, letting the curved pipe exit the head parallel to the ground, tucked up against the head. This greatly improves ground clearance, allowing the header to be mounted to the side of the head, instead of under it. One problem with the angle-tube installation is a lack of interchangeability. Spare heads for a competition car must have the same exact design for the flanges to mate. There is a new design exhaust system being offered for the street, the "Big Single." It feeds both stock manifolds to the right-hand side of the motor into a collector and uses a low-restriction, high-flow replacement muffler.

My personal preference for a street machine is for a system with original manifolds and two stock-type mufflers. This is very quiet, yet does not limit the performance potential very much. If the car is to be used in competition, header plugs can be welded into the exhaust pipes ahead of the mufflers. These openings can be uncapped when the car is raced, and closed for all other use. (Local laws may limit use of these.) Another advantage of this system is a fully functional heater/defroster system. Headers usually do a number on providing real heat to the front of the car. For a "mostly" competition machine, I have run forward-facing headers, a crossover between the sides, with a take-apart flange for removal, two big U-bends and a Flowmaster muffler on each side. This had a great sound, and still provided great flow. Turbos cannot easily use headers or dual pipes, but can still use a low-restriction muffler.

Exhaust coating can result in a long term cost savings, especially on turbocharged models. The high heat of the turbo system tends to oxidize the mild steel tubing used both before and after the turbo. After fabrication is complete, you can remove the steel items and send them off to one of several shops around the USA. Jet-Hot Coating is one such manufacturer. They apply a ceramic coating to either the outside only, or to both inside and outside of the tubing and mufflers. After baking, the finish is as tough as nails and protects the tubing or muffler from rusting out. It also maintains its good looks for a much longer time.

Be cautious in your exhaust system design and execution. With short pipes and limited room for mufflers, the Corvair can sound quite obnoxious. With many authorities mandating and enforcing noise limits, we don't need to make the Corvair stand out as the noisy one.

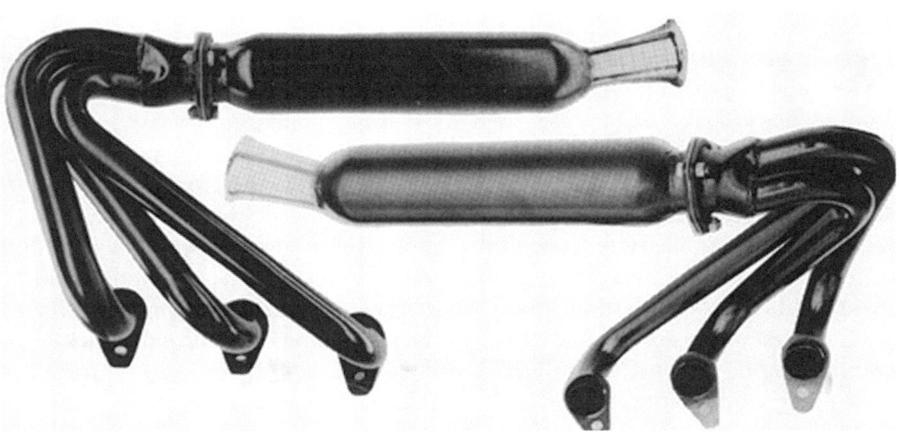

Clark's headers have the loose flange design that adapts to the head and holds the donut in place.

6

Cylinder Heads

A major stumbling block in the Corvair's performance path was created by cylinder head design. Small valves, difficult-to-reach intake-port leads, tiny exhausts, and combustion-chamber faults add up to impressive restrictions in air-pumping abilities. Even so, the construction of these heads permits extensive modifications to provide drastic pumping-capability improvements. As much as 20 HP can be added to stock engines and even more to hopped-up versions—strictly through head modifications.

Cylinder head reworking is essential to realize the full potential of the Corvair, especially with non-supercharged versions. Modifications must be made carefully and with a complete understanding of the total project. Attention to detail is an absolute must!

Simple head modifications consist of careful attention to details including raising compression, eliminating sharp edges, removing the step from the combustion chamber to increase turbulence, making all chambers of equal volume, and careful attention to valve seating. The introduction of 4 × 1 heads in 1965 brought "big-valve" heads within the reach of every serious Corvair tuner. These new heads featured 1.72"-diameter intake valves, 1.32" exhaust valves, larger ports on both the intake and exhaust sides, and added breathing ability created through the good unshrouded chamber design typical of the Chevrolet V8 heads of the time.

Major port reworking, essential for serious racing, requires cutting or milling away some or all the stock manifold to get at the intake-port leads. After porting new manifold tops and individual intake ports, new manifolds or vertical plates can be added to permit installation of any carburetion or fuel-injection system from "mild" to "wild."

Stock Head Description

Two opposing and identical cast-aluminum cylinder heads incorporate cooling fins and integral intake manifolds and contain wedge-shaped combustion chambers and valves for each cylinder. Valves are actuated by pushrods through 1 to 1.57 ratio stamped-steel rocker arms, similar to those used on other GM engines. Steel-alloy valve-seat inserts are used.

Exhaust ports are short for minimal heat transfer to the cylinder head. These connect to an exhaust manifold through short lengths of tubing pressed into the head. In addition, thick exhaust-port walls promote heat flow from the valve seat to cooling fins on the outer port wall. Wide valve spacing gives a tunnel for cooling air to pass between intake and exhaust ports. Thick sections over the chamber and surrounding the cylinder-bore spigots gives rapid heat transfer into the fins. For optimum cooling of the heads under extreme conditions, street or race, the finned areas should have all the casting flash removed from between the fins. A wide hack saw blade works well for this, as does a small Sawzall. There are also slightly thicker round blades, with cutting dust glued on. These work well, too. You can chuck up one of the round blades in a cordless drill and save some time. A couple of hours spent here might provide just the cooling improvement you will need to keep your valve seats intact.

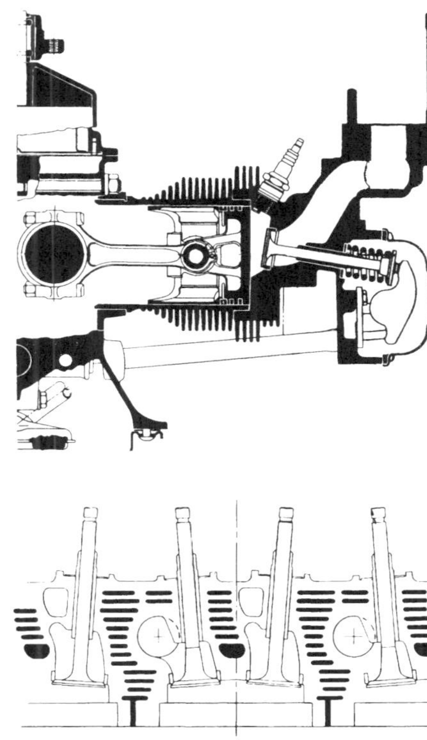

Transverse section of Corvair shows integral manifolding. Heavy black lines in head cross section indicate cooling airflow path. Be sure that these passages are fully opened for best cooling.

Guides are cast iron in all models except Spyders, which use aluminum bronze guides for improved heat removal. Exhaust and intake valve inserts are cast-chromium and cast-pickle-steel alloy, respectively. Seats are shrink fitted with interference fits from 0.002 to 0.004 inch.

The cylinder head incorporates half of the rocker-arm cover. O-ring sealed pushrod tubes inserted into the head provide for draining oil back into the sump. Mounting bosses are provided for the overheat switch, and, on 140 to 180 HP models, for a temperature sender and, for all turbo motors, a supercharger oil drain on the right side head.

Corvair's ability to use high-compression ratios is seriously limited by a "step" in the chamber that greatly reduces the turbulence that is so important in obtaining good performance from a wedge-type chamber. Even so, the chambers can be modified to provide "squish" as originally discovered by Ricardo in 1922. From P. F. Irving's *Motorcycle Engineering*, we read that the head can be machined so that the flat portion is separated from the piston crown at TDC by the thickness of the gasket. At the end of the compression stroke the mixture is squished out of the narrow gap between the piston crown and the combustion-chamber "deck" into the valve-containing recess. This creates turbulence for rapid combustion of the entire mixture charge. Because the "squish" or quench area has two relatively cool faces close together, detonation cannot occur in the unburned gasses farthest from the plug. There is a "sweet spot" amount for this squish height. Too small and the flame front can't travel inside to ignite all the fuel/air mixture, too large and there is not enough squish affect to provide mixture swirl. A distance from piston head to cylinder head of .060 to .090 inch should do it.

Computing Compression Ratios

These sketches and formulae show you how to compute compression ratios. Measuring the volumes of cylinder and gasket is easily accomplished by caliper or micrometer, or by knowing the bore and stroke. Measuring combustion-chamber volumes is more involved. A beaker or burette graduated in cubic centimeters is used because ounce graduations are not small enough to obtain the accuracy that is needed. (One ounce is 30 ccs!) The cylinder heads are leveled, plugs screwed in, and valves lightly coated with grease and dropped into place. Water can be used for making volume measurements provided the heads, valves, and guides are blown free of the water afterwards. Oil can also be used, but it is messy. If solvent is used, then the valves should be installed with springs to ensure a seal. Fill each chamber level with the gasket-register surface and note the amount required. This measurement, shown as V_3 in the sketches and formulae, is used to compute the actual compression ratio. In reworking heads, it is not unusual to make this measurement several times, making modifications between measurements to get the desired ratio.

Modeling clay can also be used to measure combustion chamber volume. With the valves in place and a flat-nosed dummy plug screwed in, the chamber and valves are wiped with a light coat of oil or Vaseline. Clay is pressed into the chamber and a straight edge used to level it with the gasket-register surface. The lump is then extracted and dropped into a cubic-centimeter beaker partially filled with water. The volume increase measured is V_3 or combustion-chamber volume.

NOTE: *Cubic inches are converted to cubic centimeters by the following formula: Cubic inches × 16.387 = cubic centimeters (cc).*

Average volume of an installed and compressed Corvair stock head gasket section is 5.4 cc. The Otto Parts big-bore gasket for '61–'64 models measures 5.7 cc. Compute the volume of other than stock gaskets by using the following formula (all dimensions in inches):

$$(\text{Gasket I.D.})^2 \times 0.7854 \times \text{Thickness} \times 16.387 = \text{Volume in ccs}$$

V_1 = Cylinder volume, assuming piston comes level with top of cylinder at TDC

V_2 = Gasket volume

V_3 = Combustion chamber volume with valves and plug installed

Compression ratio is calculated as follows:

$$\frac{(V_1 + V_2 + V_3)}{(V_2 + V_3)} = \text{Compression Ratio (Vg)}$$

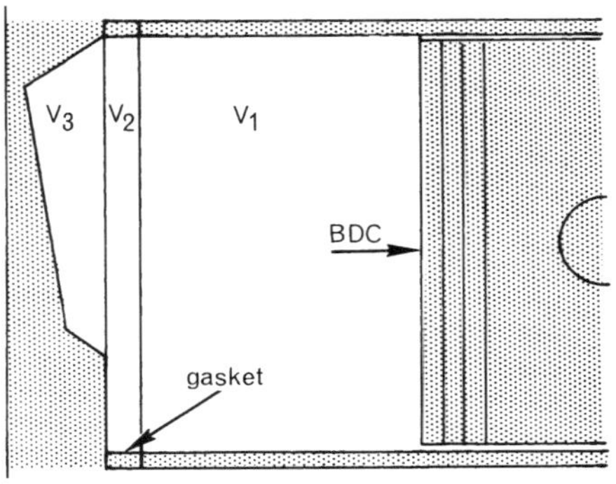

Three distinct volume areas V_1, V_2, and V_3 to consider in calculating compression ratio.

Keep in mind the relation of cylinder displacement to compression ratio. Anything you do to increase V_1 automatically increases the Vg. If you plan to rework your cylinder heads now and later bore or stroke the engine, then use the ultimate displacement to determine the volume of the head chambers now so as to avoid reworking the heads twice. A 170 cu. in. Corvair needs a 49 cc chamber for 9.5:1; a 176 cu. in. engine needs a 51 cc chamber for 9.5:1.

Simple Modifications

Even "simple" modifications to the Corvair head appear somewhat involved when all of the details are considered. The following items are required: valve-facing and seating equipment, hand or flexible-shaft grinder, piloted cutter to open up diameter of valve-spring seats on 80 HP heads, and access to a milling machine operated by a competent machinist.

First, the head must be disassembled and completely cleaned. Removing carbon from the combustion chambers takes time and care. Although a wire brush in a drill motor is sometimes recommended, it must be a fine wire brush similar to that supplied by Kent-Moore as No. J-8358. The best results are obtained by leaving the gasket in place to protect the important seating surface and using an old set of valves to protect the seats.

Another method of removing carbon deposits is to soak the heads in your 4-gallon can of carburetor cleaner that you should always purchase before working on a Corvair engine.

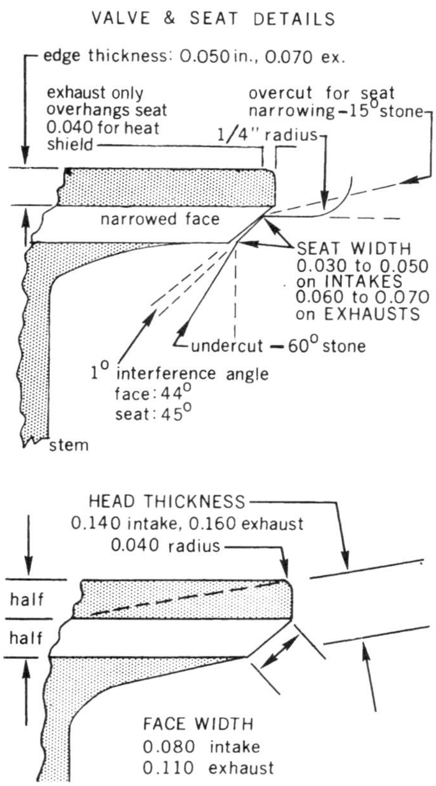

The two valve and seat drawings, supplied by OTTO Parts, detail important information you need to do a high-performance valve job. Attention to these details is worth 10 HP on the Corvair, and valve life is increased simultaneously. They also recommend that the intake valves be seated on the outer third of their faces, exhausts in the center of the valve faces. Seat run out must not exceed 0.002 inch as measured with a valve-seat dial indicator.

Overnight soaking of the heads renders the carbon deposits soft enough so they can almost be wiped out with a rag. It definitely loosens the carbon from the aluminum to make cleaning much easier. Or, carbon deposits can be quickly removed by walnut-hull blasting. This differs from other forms of cleaning in that most parts need not be masked to protect their machined surfaces, even though carbon is completely cleaned away by the process. Large aircraft-engine shops usually have such equipment. Sandblasting, another cleaning method, is useful if the sealing surfaces are protected with masking tape.

After disassembly and cleaning, the valve guides should be checked for wear as described in the Shop Manual. Excessively worn guides should be replaced or reamed and oversize valves installed. These operations are also described in the Shop Manuals. 1964 was the first year in which Chevy recommended replacement guides, even though replacement-part makers have made them since 1961. Chevrolet also listed oversize guides as well as oversize-stemmed valves to take care of any problem that could arise in head rebuilding. While Chevrolet does not sell replacement guides for the Spyder engines, several vendors have them at reasonable prices.

Any Corvair with 10,000 miles on it could use a valve job to sharpen up its performance. But, not all valve jobs are alike, and again attention to details pays off in added performance. Up to five additional HP can be gained from the stock engine by a valve job that takes full advantage of the valve size. The seats must be moved outward to coincide with the intake-valve edges, and seats narrowed to 0.060 inch on the exhausts and 0.030 inch on intakes.

Sharp edges are not wanted in combustion chambers because they cause detonation. Detonation is destructive knocking resulting from rapid and uncontrolled ignition of the fuel mixture at the wrong time during the engine's operating cycle. Therefore, you must eliminate sharp edges wherever they occur in the combustion chamber. Valve edges often require radiusing, and chamber edges at the gasket-register surface are another offender. Plug threads that are exposed with the plug fully torqued into place are prime troublemakers. If threads of the plug boss are exposed, grind back the exposed threads so the plug end comes flush with the chamber when torqued in place. Attention to these details is vital to your engine's longevity!

In general, the average dual-purpose Corvair, or even cars used solely for transportation, should be equipped with a true compression ratio of 9:1 to 9.5:1. Premium gasoline is required by these ratios. Higher ratios are tolerated by engines using racing cams because these reduce the low-RPM pumping capabilities and the higher compression helps to gain back some of the low-speed torque loss. Spyder engines (supercharged) should be restricted to true compression ratios not higher than 8:1.

Compression-ratio measurements on hundreds of stock Corvairs have shown beyond doubt that advertised ratios are higher than actual. This error is usually about one full ratio, that is, an advertised 9.25:1 often measures out at only 8.25:1. Interestingly enough, factory-specification sheets have shown two different chamber volumes for engines advertising the same compression ratio (with same displacement). For example: 1964 base engine ratio is advertised at 8.25:1, volume is 3.795 cu. in.; 1964 Spyder engine ratio is advertised at 8:1, volume is 3.919 cu. in. The base engine computes to 7.6:1; the Spyder at 7.4:1 c.r.

When raising the compression, as you would ordinarily, to get the engine up to advertised specifications, or higher, you would usually consider milling the heads. On most cylinder heads a single mill cut with a simple setup is all that is required. Corvair heads, on the other hand, are not simple to mill, even though this operation is essential to raise the compression to the desired value and to correct the chamber design for high performance.

Head Milling

Head milling is practiced for two important reasons: (1) to raise the compression ratio, and (2) to increase turbulence through reduction of the deck height and elimination of the "step" in combustion chamber squish areas. When milling Corvair heads, it is absolutely essential that the machinist fully understands that gasket register

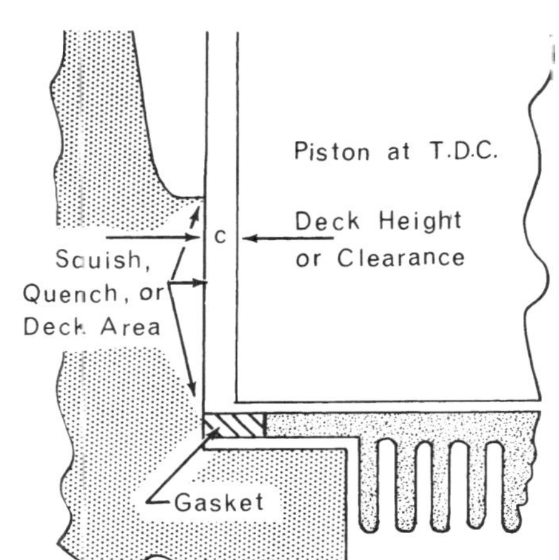

EARLY HEAD COMPRESSION RATIO (all except 140)

Calculations include gasket volume of 5.4 cc.

ENGINE MODEL (modification) BORE × STROKE	Total / One cylinder displacement (cubic inches)	Cylinder Head Type (see explanation below)							
		1	2	3	4	5	6	7	8
1960 (stock) 3.375 × 2.6	140 / 23.3	7.3	7.1	7.9	7.1	7.8	6.6	7.3	6.5
1960 (stock bore, '64 crank) 3.375 × 2.94	158 / 26.3	8.2	7.9	8.9	7.9	8.6	7.4	8.1	7.2
1961–63 (or '60 bored 0.060 inch) 3.437 × 2.6	145 / 24.1	7.6	7.4	8.2	7.3	8.0	6.9	7.5	6.7
1961–63 (0.125 inch overbore) 3.562 × 2.6	155 / 25.9	8.1	7.8	8.7	7.8	8.5	7.3	8.0	7.1
1961–64 (1964 crankshaft) 3.437 × 2.94	164 / 27.3	8.4	8.2	9.1	8.1	8.9	7.6	8.3	7.4
1961–64 (0.125 inch overbore,'64 crank) 3.562 × 2.94	176 / 29.3	9.0	8.7	9.7	8.7	9.5	8.1	8.9	7.9

Cylinder Head Specifications (volume of one chamber in cubic inches and stock compression ratio), advertised c.r.

1 1960 Head, 3.34, 8:1
2 1961 Head, 3.46, 8:1
3 1962 Head, 3.02, 9:1 (102 HP)
4 1963 Head, 3.5, 8:1 (80 HP or SPYDER)
5 1963 Head, 3.11, 9:1 (84 or 102 HP)
6 1964 Head, 3.79, 8.25:1 (95 HP)
7 1964 Head, 3.39, 9.25:1 (110 HP)
8 1964 Head, 3.92, 8:1 (SPYDER)

surfaces are to be machined exactly the same amount and that an equivalent amount must be milled from the head surface which surrounds the combustion chambers. Be careful not to cut into the sparkplug threads. Also, remember that increasing the displacement increases the compression.

As you can see from the Early Head Compression Ratio table, a list of dimensions to machine from a Corvair head would be meaningless. Furthermore, best performance results from completely eliminating the step in the combustion chamber to provide deck clearance (height) approaching the optimum of 0.025 inch.

When this machining has been accomplished, in all six chambers and on Surface S of both heads, the chamber volumes can be measured and the compression ratio calculated. Additional volume to reduce the c.r. to the desired value can be added by (1) thinning the valve heads, (2) straightening the chamber at the wedge edge, (3) grinding away the chamber to coincide with the gasket, (4) removing metal from around the plug boss, and (5) by milling a step into the crown of the piston under the non-squish area of the chamber.

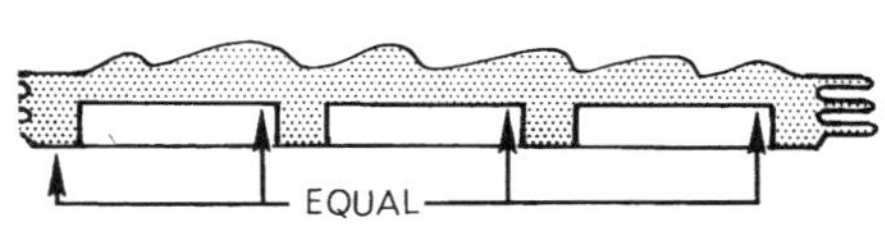

All cylinder-register surfaces must be milled to exactly the same depth to ensure gasket seating. The amount milled from register (gasket) surface must also be milled from the face of the head (surface S) to avoid interference with the top fins on the cylinders.

Polishing the valve heads and combustion chamber surfaces, as well as the piston crowns, reduces the heat loss to these components, thereby improving power output. Polishing provides a

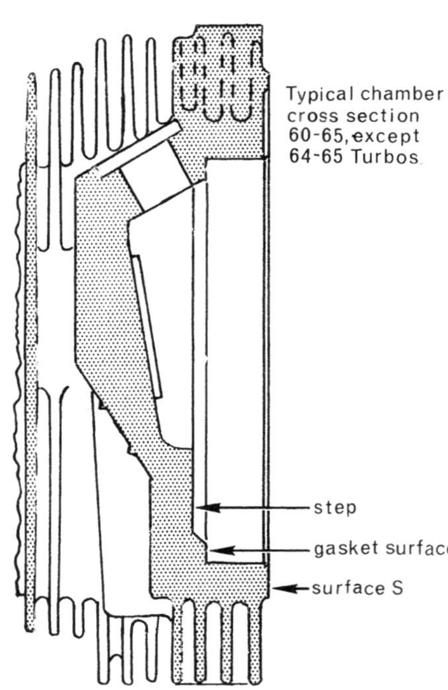

1965-66 140 HP HEAD WITH ONE STANDARD (.018) UNDER-BARREL GASKET

Calculations and table provided by Bert Hultman. • All dimensions given in inches or cubic inches.

Overbore on Stock Cylinder Barrel

Requires Special Cylinder Barrels

		Stock Bore/ Stroke	+0.010	+0.020	+0.030	+0.040	+0.060	3-9/16	3-5/8
Bore/Stroke (In)		3.437/2.94	3.447/2.94	3.457/2.94	3.467/2.94	3.477/2.94	3.497/2.94	3.562/2.94	3.625/2.94
Head Chamber Volume	Stock	3.390	3.390	3.390	3.390	3.390	3.390	3.390	3.390
	+.080 Mill	2.648	2.648	2.648	2.648	2.648	2.648	2.648	2.648
Displacement	Total	164.00	164.61	165.57	166.53	167.49	169.42	175.83	182.05
Cubic Inches	Each Cylinder	27.333	27.436	27.595	27.755	27.915	28.237	29.305	30.342
	Head Gasket Thickness	***Compression Ratio (Stock Advertised 9.25:1)***							
Unmilled Stock Head	.032 (std)	8.413	8.438	8.477	8.517	8.557	8.637	8.902	9.156
	.042	8.231	8.254	8.292	8.329	8.368	8.444	8.694	8.936
	.052	8.058	8.079	8.115	8.151	8.187	8.260	8.498	8.727
	.064	7.861	7.881	7.915	7.948	7.982	8.051	8.275	8.491
	.074	7.705	7.723	7.755	7.788	7.820	7.886	8.099	8.466
Head Milled .080	.032 (std)	10.281	10.311	10.359	10.408	10.456	10.582	10.877	11.188
	.042	9.998	10.025	10.071	10.116	10.162	10.254	10.556	10.846
	.052	9.731	9.756	9.799	9.842	9.885	9.971	10.255	10.527
	.064	9.431	9.454	9.494	9.534	9.574	9.654	9.918	10.171
	.074	9.197	9.218	9.255	9.293	9.331	9.407	9.665	9.893

To convert cubic inches into cubic centimeters (ccs) multiply by 16.387. Example: (Stock) 3.390 cu. In. × 16.387 = 55.55 cc.
To convert cubic centimeters into cubic inches (ccs) multiply by .0610.

secondary benefit by making carbon removal easier when the engine is taken apart for cleaning or a valve job.

If you have obtained the idea that head work of even the "simple" variety is really a lot of hard work, you are 100% correct. Corvair cylinder-head reworking, or almost any Corvair work for that matter, cannot be given to just any "schlock-shop" mechanic if performance and reliability are required. The mechanic and the machinist must have a complete understanding and appreciation of what work is being done. In order to build a serious competition motor out of core material at least 50 years old requires a thorough evaluation of the hardware. Most race motors are built out of older 140 heads. To reach the level of reliability needed to win races, those heads have all the valve seat inserts removed and replaced by "deep seat" inserts with either major shrinking with heat and cold or physical material displacement over a stepped outer design. Also, the valve guides have been replaced and new guides "tailored" for the installation. Lots of chamber and port work encourage the production of more horsepower.

Head milling raises the compression ratio for more HP. At the same time, turbulence is increased by improving cylinder-head design, improving the engine's ability to use the higher compression. A point not often mentioned is gasket-surface deformation that occurs on '60–'64 Corvairs due to the narrow-width head gaskets that are used. Milling the heads removes the deformation to restore a perfect gasket seating and sealing surface, but subsequent installation of the same heads may require cleaning up this surface again. To mill the heads, position one head on a milling-machine table, centering one chamber under a cutter adjusted to cut to spigot diameter. This cutter should have a 1/32-inch radius at its tip. On non-Spyder engines, cut down the gasket-sealing surface approximately 0.110 inch or until the "step" is eliminated. Record the depth of the cut. Lift the cutter out of the chamber bore and dial over 4.80 inch to locate the same operation in the next chamber. Mill each of the two remaining chambers the exact depth used on the first chamber. Then machine the other cylinder head exactly as you did the first one, After finishing all six of the chambers, mill Surface S the same amount which you lowered the gasket surfaces. If you milled the gasket surface 0.110 inch, take that much off of Surface S to permit the cylinders to enter the heads to compress the gaskets. Otherwise, the top fin on each cylinder will foul the head to prevent gasket sealing. Be sure that the machinist understands the absolute need for equal-depth cuts in each chamber, or blown gaskets will result.

Milling Corvair heads usually requires pushrod changes as detailed in the "Camshafts and Valve Train Basics" chapter. Failure to ensure that correct valve-to-rocker-arm geometry is established will seriously reduce the performance by subtracting from specification valve lift. The careful tuner must not overlook this seemingly insignificant point. Also, milling Corvair heads moves them inboard toward the center of the engine. As a result, the main blower shroud will not fit until you file the holes inward to meet with the relocated cylinder-head bosses. Cutouts around the carburetor mounting pads must also be trimmed away for clearance, but this should be done with extreme care to keep cooling-air losses to an absolute minimum. And, the generator-bracket holes will also need filing to fit.

Finishing Operations

Use machinist blue dye to coat the gasket-sealing surfaces in each of the chambers. Drop gaskets (big-bore or stock, depending on bore size to be used) into each head opening. Scribe a line around the I.D. of each gasket. Note that there is metal between the gasket I.D. and the valve edges that are nearest to the gasket. This metal should be removed to within 1/32 inch of the scribed line to "unshroud" the valves for better breathing. Removing metal from the chamber reduces the compression ratio, but tests have proved that improved breathing gains more HP than the reduction in compression loses.

A die grinder with a rotary file is the best tool to use for cylinder-head work, The same tool can be used for removing sharp edges, grinding back exposed sparkplug threads, and when sanding sleeves are used, the tool is also perfect for polishing the chambers. Be careful that you do not nick the valve-seat surfaces. Aluminum removal is speeded by keeping the file free of chips. An occasional application of a bar of soap to the spinning rotary file will lubricate it and keep it chip-free.

Grind the valve seats as illustrated in the drawings. Reinstall the valves. With a graduate or burette measured in cubic centimeters, measure and record the volume of each chamber. When you have measured all of the cylinders, find the one with the greatest volume. Calculate the compression ratio that this volume will provide. If it is the desired ratio, increase all the remaining five chambers to the same volume, checking and rechecking as you remove metal. If the ratio is too high, then additional metal will have to be removed from the largest chamber until you get it to the desired figure, then modify all of the other chambers to the same volume. Polish the chambers.

The entry path of the ports under the valve seats can be smoothed up somewhat, but do not reduce the length of the valve guides. Polishing of the intake ports does very little good where performance is concerned, so do not waste your time in this area.

Head Gaskets

Special head gaskets are required on reworked Corvair heads for two reasons: (1) to reduce deck clearance to a figure close to the desirable minimum, and (2) so the gasket will not hang into the combustion chamber and bore to glow red hot. Stock gaskets will extend into the combustion area on any bore size larger than 3-7/16 inch + 0.020 inch. The use of stock gaskets in a modified engine often causes "running-on" and destructive pre-ignition. Big-bore head gaskets are made for '60–'64 heads to fit bore sizes to 3-9/16 inch. The same gaskets can be used with the early-style cylinders (pre-1965) used with the 1965 4 × 1 heads. Instructions are included with the gaskets.

Thor die grinder is a perfect tool for working on Corvair and other cylinder heads. Three collet sizes permit using a wide range of readily available rotary files and sanding sleeves to eliminate sharp edges, grind away exposed plug threads, and equalize combustion-chamber sizes.

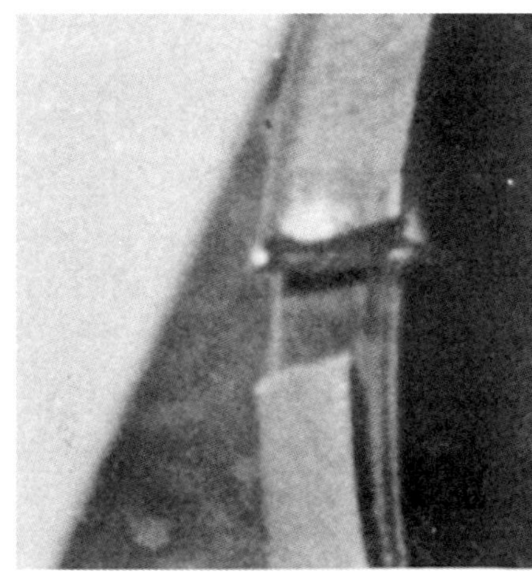

Comparison of stock and big-bore gasket illustrates need for using the special gaskets on any overbored Corvair cylinder. Stock gasket hangs into cylinder, can glow red hot to cause detonation. In some instances, stock gaskets interfere with oversize pistons.

How to Get Big Intake Ports

Openings from the manifold "log" into the intake-port leads are restricted, as shown in photo 1. Yet, by milling away the stock manifold, the ports can be opened up considerably for improved breathing capabilities (photo 2), a plate or special manifold or individual intake ports can be applied to the modified head. Plates or special manifolds are bolted on, whereas the individual port leads are usually heliarced into place. Port openings of the milled-off heads can be made rectangular with rounded corners. The opening size will be approximately 1-¼ × 1-½ inches (photo 3). Do not grind away the guides to become level with the port. Doing so spoils the centering of the valve on its seat, makes your careful seat jobs last fewer miles, and increases oil consumption through the guides. No real gain can be obtained by cutting off the guides.

1

2

3

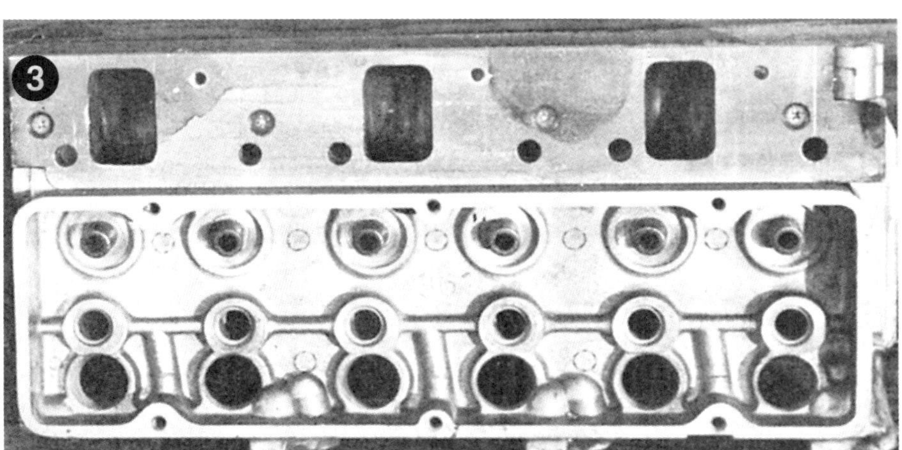

140 HP 4 × 1 Heads

The heads introduced on the 140 HP four-carburetor engines for 1965 can be installed without modification for drastic improvements in the performance of any Corvair engine. They have larger ports, provisions for mounting four carburetors (two on each head), and 1.72-inch intakes with 1.36-inch exhausts. In addition, larger exhaust tubes carry the combustion products away from the exhaust valves. A complete set of these heads with valves, two extra carburetors, linkage, and oversize exhaust manifolds costs less than prices ordinarily charged for modifying two Corvair cylinder heads to include big valves. As is the case with all 1965 heads, the bores for the cylinders are larger to accommodate wider head gaskets, Using these heads with 1960–64 cylinders requires inserting a piece of 0.025-inch shim stock (19⁄32 inch wide by 11-29⁄32 inch long) around the inside of each cylinder bore in each cylinder head. As is the case with any other Corvair head, special attention must be given to the elimination of sharp edges, exposed sparkplug threads, and carefully seating the valves. Removing the step from the chamber and higher compression is also recommended, together with equalizing the chambers. In the stock 140 HP intake manifold, the sharp edge at the bottom of the drilled carburetor port can be radiused and smoothed with a reverse tapered cutter. This will help the mixture entry into the log and into the chamber. For all-out competition with a 4-one-barrel setup, you can cut the manifold off the head by standing the head up and band-sawing the manifold off at an angle, then porting the actual routes into the chambers, and welding the manifold back on when you are finished porting. Angle cutting allows you to reach all needed locations for rewelding. This extensive and expensive mod would only be done when stock carburetor mounting is required by the rules. Just as a reference, a 140 HP head can have the secondary mounting pad removed and plug welded to provide the appearance of a 110 cylinder head, utilizing 110 engine shrouding.

Comparison photographs of small valve head (above) and 140 HP head (below). Difference in valve sizes is indicated by the material between the seats.

140 HP exhaust tubes have a much larger cross section.

Double the intake capacity for additional carburetion.

The 140 HP heads, Part No. 3878565, were only $60 each, new; intake valves, Part No. 3856734 were $2.42 each; exhausts, Part No. 3856735, $4.60 each; and stock-bore head gaskets, Part No. 3856720. Many different replacement valves are now available in the aftermarket, including stainless steel construction.

Because each head has two carburetor-mounting pads, modifications to the blower shroud are required. Because the exhaust tubes on the heads are larger, you can increase the openings in your exhaust manifolds to 1-$^{21}/_{64}$ inch. Or you may prefer to buy the 140 exhaust manifolds that have a larger cross-sectional area for improved breathing. Part No. 3856741, new

Combustion chamber of 140 HP head for 1965 is similar to Chevy V8. Round any sharp edges, especially around plug threads, before installation. Step can be removed to increase turbulence.

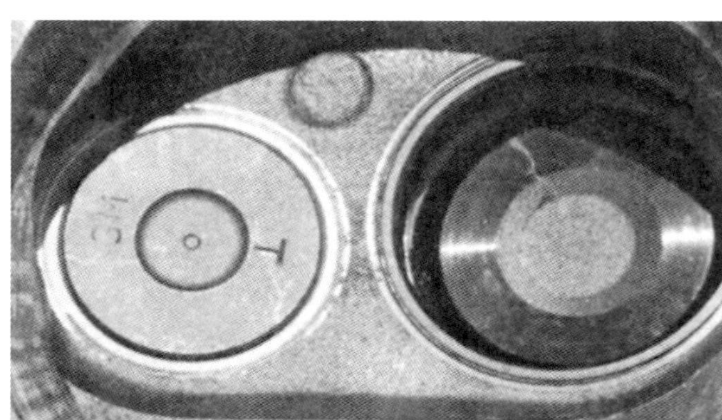

Photo shows same chamber with smaller Corvair valves dropped into the guides.

replacements are available from Clark's. The heads can be used with either two or four carburetors. Two-carb use merely requires blocking off one opening on each head. The heads are also helpful when installed on Spyders, but exhaust systems will be custom.

Three-Barrel Weber Carbs

Perhaps the most impressive cylinder head modification available is the conversion to a pair of Weber 3-barrel carburetors. It involves some major machining to the Corvair head and some touch-up TIG welding of the head (see photo 3 page 47). Although known mostly as a Porsche 911 carb, the Weber IDA3C was originally designed for a small Alfa V6 in the 1950s. They are still available new, though quite "pricey." There are other 3-barrels produced as replacements. They do turn up at swap meets, usually Porsche swap meets, sometimes at reasonable prices. Before you put down a few hundred dollars for a pair of someone else's castoffs, check thoroughly for corrosion. The Weber uses a cast aluminum body and the float bowls don't like water. Many brass items screw into the housings, such as jets and covers. You will need to remove them for a rebuild and corrosion can make that quite a chore. The conversion manifolds that adapt the Webers to the Corvair heads have been made by at least four companies. Many of the original Crown manifolds still float around the market. Corvair Connection built them for a number of years and SC Performance may still be offering kits (and the machining of your heads on an exchange basis). Linkage has always been a sore point for the dual Weber installation as the carbs are offset fore and aft from each other.

The Corvair Connection made several improvements on the Crown kit. Gaskets replace O-rings. Manifold flanges were drilled and tapped for linkage mounting. Drain holes were added to protect linkage from corrosion. Any Weber kit will need cylinder head flange work to help smooth the mixture flow from the manifold to the head.

7

Camshafts and Valve Train Basics

That the camshaft is the major performance-determining element in the Corvair can be seen from studying the engine-performance graphs for the various years when the car was manufactured. Engines with identical displacement, carburetion, and valve sizes have outputs varying by 22 advertised HP. The figure becomes even larger when net HP differences are compared. Often the only major changes that you will find in engine specs are in the camshaft department.

Because the camshaft is the last thing that falls out of the Corvair engine during disassembly, it is also the last item that is usually changed when chasing after additional power. By the same token, it needs very careful installation to ensure that maximum benefit is obtained from its action. The camshaft is only part of the valve train and the other items deserve equal attention if utmost performance is to be achieved. Hydraulic lifters, pushrods, rocker arms, spring retainers, and valve springs interact in ways that are not completely understood by many mechanics. Installation techniques are also important because they affect camshaft life. The coverage devoted to these related parts and to careful installation may seem overly long to the casual reader, but the devoted engine builder will recognize that there is no real way to shorten the presentation and retain the complete picture.

Several camshafts have been offered in stock Corvair engines. Due to the manner in which Chevrolet states valve timing, which includes lengthy clearance ramps that contribute nothing to performance, looking at factory specs would give you the idea that grinders of special cams can do little to help. Such is not the case! A detailed discussion of effective valve opening, as well as timing graphs relating stock cams to reground ones, should aid in dispelling some of the hocus-pocus surrounding the entire subject. Valve spring and retainer information is also included, together with specific directions for checking and changing camshaft timing at installation. The coverage of rocker-arm geometry and its effect on valve lift is the most complete that has ever been presented to this time. The reader who follows all of this information with great care will discover that his engine has suddenly gained free horsepower—and that is the best kind!

Basic Facts About Camshaft Functioning

Although it looks like a simple stick with bumps on it, the cam is the major controlling element in determining your engine's ultimate behavior. The stock camshaft is carefully related to a fantastic number of factors, including carburetor area, compression, displacement, transmission and axle ratios, car weight, performance desires, driving habits of the U.S. buying public, and so on ad infinitum. You and I cannot relate these factors to the same fine degree as do the factory engineers with their experience and computers, but we can consider what goes on during the intake and exhaust phases of a single cylinder.

Valve timing is nearly always related to degrees of crankshaft rotation, but this oversimplification should be avoided. Another relationship is more helpful in describing camshaft functioning: valve action versus displacement of the cylinder. Where is the piston and what is it doing?

As high school students, we were taught four-stroke internal-combustion engine functioning in four unrelated 180-degree segments: Intake, Compression, Power, and Exhaust. If the valve mechanism could slam the valves open and shut precisely at the TDC and BDC points without delays, the engine would probably run until some part broke! But, just as instantaneous valve action is impossible, due to the severe forces that would impose on the valve mechanism, instantaneous movement of the intake and exhaust gases is also impossible. So, the 180 degree-segmented description of valve opening versus crankshaft position does not work outside of the textbook covers. Now that we have shattered our schoolboy beliefs, let's look at the practical facts of the matter.

Basically, the intake valve is opened before TDC while the piston is still rising on the exhaust stroke. There is not much pressure left in the cylinder at this time and the early opening is added to the textbook's 180 degrees, even on "mild" engines. Early intake-valve opening on the Corvair gives the valve a head start on the movement of the piston. The valve must be limited in its rate of acceleration so the cam designer starts it moving early to cause it to be fully open when it can do the most good. It must be fully open before the piston reaches its point of maximum acceleration or the gas flow will be impeded by a partially open valve just as cylinder displacement is being increased at the most rapid rate. Early opening is often touted as a cleaning device for pushing the exhaust out and helping to clean up the chamber of spent gases. According to Racer Brown, this may be true of a hemispherical chamber with a tuned exhaust and intake system, but it certainly does not apply to the Corvair's wedge-shaped chamber. Early opening in the Corvair should be approached with great care because the incoming gas has a decided tendency to do an "about-face escape act" through the convenient exhaust valve opening. All you get for your efforts is increased fuel consumption, plus a few other disadvantages, such as increased emissions and perhaps a bit of exhaust-valve cooling by the escaping fresh mixture. As you increase the number of degrees that the intake valve opens ahead of TDC, the intake system is confused at low speeds. Early opening, therefore, is the culprit causing the rough-idle trademark of a high-RPM engine. While early opening causes no particular problems at high speeds that provide an atmospheric-pressure operating condition, reduced speeds cause manifold vacuum to suck exhaust gases into the intake system. The results are diluted fuel/air mixtures, rough idling, sparkplug fouling in extreme cases, and fogging of the mixture out of the carburetor or injector throats. This dilution is one of the reasons why an accelerating racing engine will tolerate full ignition advance at very-low RPM.

Just as we added to the opening side of the 180, the same must be done at the finish. There is good reason to leave the valve open after BDC to take advantage of the incoming-gas momentum. Although the piston can be said to be starting its upward stroke after BDC, the truth of the matter is that it is barely moving upward for quite a few degrees. The crank swings through a considerable arc to swing the connecting rod big end sideways. This is especially true for Corvair with its short connecting rods. The momentum of the inflowing gases gives additional filling during this period of piston "laziness," up to the point where reverse pumping begins. As piston-acceleration rate increases, the intake valve must be closed or severe reverse pumping and charge-density reduction will occur. It should be obvious that as engine RPM is increased, the closing point can be left until later than is desirable on an engine which has to produce good torque in the low and mid range. A tuned-intake system will increase charging-mixture momentum to improve cylinder filling in a limited RPM range. Reverse-pumping of the charge back into the manifold, as caused by late closing, is one of the factors allowing modern day super-stock engines to be set up for drag race competition with much higher compression than would ordinarily be possible with gasoline for fuel. The same cars would require lower compression if set up for continuous high-RPM operation, unless running at high altitude, as at Bonneville.

Exhaust-valve operation can almost be deduced from what you have read about the intake. Because the exhaust valve needs its "head start" on the piston, a part of the power stroke is subtracted from the 180 example by opening the valve before BDC. This releases pressure from the cylinder so the piston works less to pump out the gases, and ensures that the valve is fully open by the time when the piston is accelerating at its maximum upward rate. Pressure "blow-down" also starts the exhaust gas flow out of the cylinder. Strangely enough, little power is lost by virtue of early exhaust-valve opening because the burning gases impart most of their available effort to the piston in the first 90 degrees of the power stroke. For this reason, some cams are ground with dual patterns to give more exhaust than intake duration. When closing the exhaust valve, the intake action is overlapped—both valves are open at the same time. If this overlap is stretched too far, serious dilution occurs, especially past 70 degrees overlap. When a tuned-exhaust system is used, late closing of the exhaust takes advantage of the momentum of the leaving gases to help discharge the cylinder. It is sometimes possible to reduce cylinder pressure on a momentary basis to below atmospheric (vacuum) so the incoming gases are started in by the draft of the exhaust gases, as well as by the increasing displacement afforded by the descending piston.

While the foregoing can be interesting in understanding the operation of your Corvair engine, it would not enable you to tell the cam grinder the precise number of degrees of overlap and duration that you would need for specific applications. Fortunately for most of us, the cam grinders use their superior knowledge of mechanical limitations and the physics of gas flow to protect us from ourselves. They have cam patterns for every situation. You cannot go far wrong by relying on the recommendations of qualified and capable cam grinders, especially if you are willing to follow through with careful installation procedures as will be described.

Corvair's valve train is nearly identical to that used since 1955 in Chevy and Pontiac V8s and 1962–64 Ford 221, 260, and 289 engines. The design has been thoroughly proved. An aluminum gear-driven camshaft of alloy iron with nine lobes (exhaust lobes common for opposing cylinder pairs) runs directly on bores in the aluminum crankcase halves. Hydraulic lifters also operate in holes bored in the aluminum cases. Tubular pushrods are made of 5/16-inch diameter, 1/32-inch wall steel and are approximately 10 inches long. These reach through rubber-ring-sealed oil return tubes to connect with rocker arms in the "tub" shaped section atop each head. Each rocker arm is formed from pressed steel and pivots on a half-ball centered on a stud. Rocker adjustment is by self-locking nut. Unlike the V8, the Corvair rocker has to make up for an alignment difference between the ball studs and the valves. The V8 valves are parallel. The Corvair valves are splayed. This means the Corvair rocker arm has to be able to "lean-over" a bit during operation, so the slot in the bottom of the rocker is wider than the V8. The lower portion of the rocker stud also functions as the lower head-attachment nut. Camshaft motion is multiplied approximately 1.57 times by the rocker arm that contacts the tip of an intake or an exhaust valve (1.34-inch and 1.24-inch diameter, respectively, larger on the 140 HP). Valve stems are $\frac{11}{32}$-inch diameter. Seating pressure for each valve is provided by a coil spring 0.78-inch or 0.88-inch diameter on 1960–63 models. The smaller springs are used on 80 HP models and the larger springs with flat-wound damper coils are standard on all 1964-up models. Springs are retained by stamped-steel retainers weighing 23 grams each and wedged to the valve stem by one of two valve-key styles. Retainers are made in two diameters to fit the two spring sizes: 1-$\frac{1}{32}$ inch and 1-$\frac{3}{16}$ inch.

The valve mechanism is sealed off by a stamped-steel cover that forms the outer half of the enclosure. Constantly bathed in a thin film of oil, these covers are a significant aid in cooling the Corvair engine.

Stock Corvair Camshafts, GM Specification Data

The following table indicates year, intake opening and closing, exhaust opening and closing, and lift in inches. Opening and closing figures are in degrees of crankshaft movement.

	Standard	High-Performance	Turbocharged
1960	29–65 73–41** 0.314 0.344	45–80 81–54 0.380	
1961	Same as above	Same as above	
1962	Same as above	Same as above	Same as High-Perf.
1963	Same as above	36–62 72–22** 0.380	58–86 98–46 0.375
1964–1967	29–60 60–80** 0.380 (1)	37–81 79–39** 0.391	Same as High-Perf.

(1) The '64-'67 standard cam is also used in high-performance engines when they are installed into Powerglide transmissions. The four asterisked (**) camshafts were available in the '67 Chevy Parts Book at approximately $18 each. Current hi-performance part 3872304 is a terrific cam for the street and can be considered an excellent value for almost any Corvair owner. Although the listed timings are adequately "wild" for some competitive use, the camgrinders should always be consulted as they have plenty of tricks up their sleeves and the effective opening and closing points of the same cams tell a different story when compared with special ones.

Effective Timing Comparison: Stock vs. Special Camshaft

The following table compares three stock Corvair cams with three of the popular reground ones all on the basis of effective timing, that is with the opening and closing measured with the valves 0.015 inch off their seats.

	Intake opens/ closes	Exhaust opens/closes	Lift	Duration	Overlap
1960–63 Stock 80 HP	3–30	46–6	0.314	213° intake 233° exhaust	9°
1960–62 Stock 95, 98, 102 HP	20–54	62–13	0.380	255°	33°
1964–65 Stock 110, 140, 150, & 180 HP	21–60	64–18	0.391	261°	39°
Racer Brown SS H-9	29–57	57–29	0.424	266°	58°
SS H-25	Good for competition only or for blown engines, recommended only for knowledgable tuners, 0.485 lift, 286° duration.				
SS H-30	30–62	66–26	0.450	272°	58°

Racing Cams

Perhaps the best way to start examining the subject of racing cams is to relate a conversation with W. G. "Racer" Brown. Well known for his Porsche records at the Bonneville National Speed Trials, Racer has transferred this expensively obtained air-cooled-engine experience to the Corvair. He says, "One cannot go mad on valve timing on the Corvair, primarily because of small cylinder displacement and also because of manifolding problems that abound in stock or modified-stock versions. It is therefore advisable to keep effective duration about 280 or 285 degrees as a reasonable maximum, with around 260 degrees or so for street operation, particularly on the inlet side. The exhaust can stand additional duration that will not detract from overall performance. Generally, valve lift should be in the range of 0.400 to 0.450 inch. Higher lifts will not show up as well as on other engines because the Corvair valves are pretty well free of combustion chamber walls, adjacent valves, etc. In other words, entry and exit to and from the combustion chamber is no particular problem. Upstream of the inlet valve is where the real problem lies.

Hydraulic tappets are essential in this engine because of rather wide temperature differentials under normal operation which combine with thermal expansion and/or contraction to cause relatively large valve-clearance

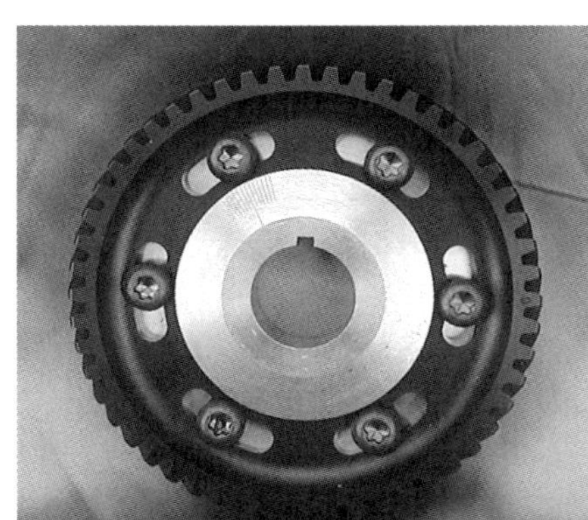

This adjustable cam gear from American Pi allows the cam to be advanced or retarded after installation. It does require removal of the bellhousing to do the change, but that sure beats splitting the cases.

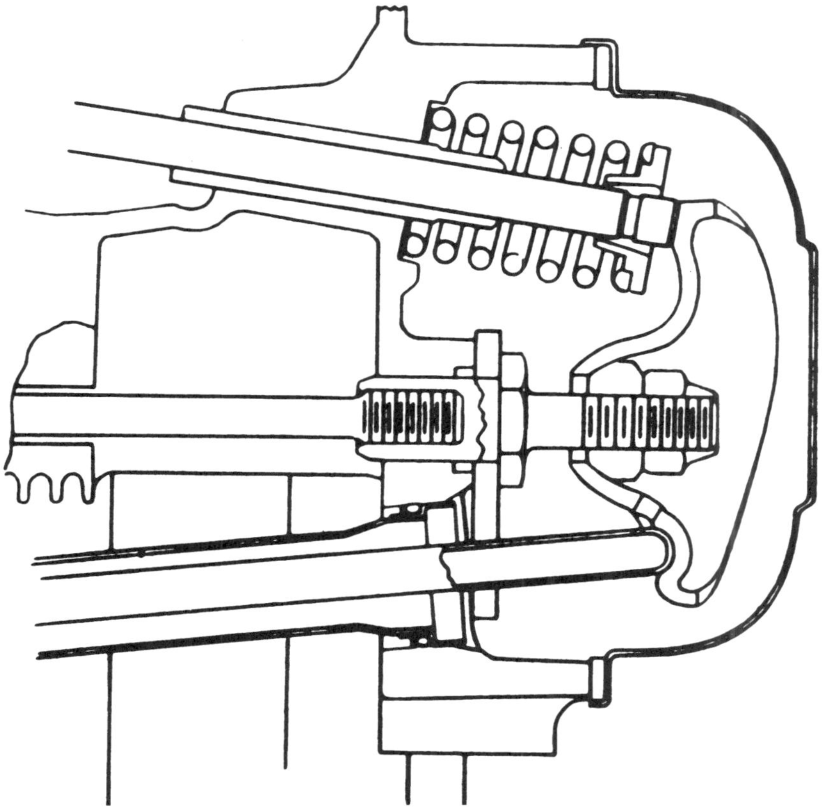

Cross-sectional drawing of valve train at the cylinder head end is from original SAE Paper, *The Chevrolet Corvair.* O-ring seal on head-nut/rocker-stud not shown.

Rocker cover oil leaks were "dried up" with important sealing changes in 1964. Owners of early Corvairs can obtain the improved parts from Chevy dealers.

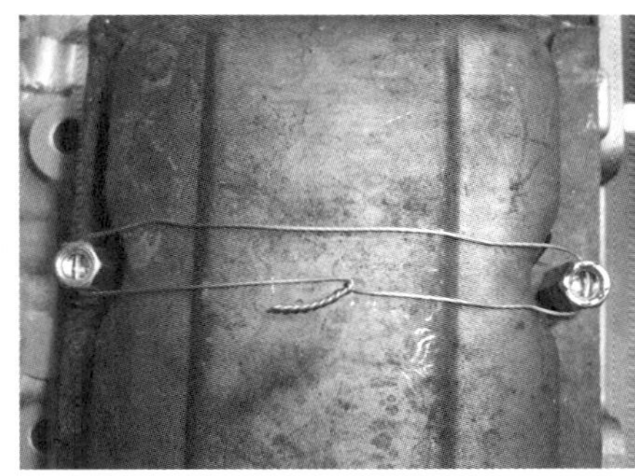

This system will assure that your Corvair engine will never leak oil from the valve covers. Studs, drilled nuts, and a little safety wire.

variations with mechanical-lifter camshafts."

Racer continued, "There is no effective timing until the valve is off its seat at least 0.015 inch to establish flow. It is easy to get added duration for advertising purposes by adding in the lengthy ramps that must be included to lift the cam gently off of its seat and return it thereto in the same fashion. When the effective-timing yardstick is applied, stock Corvair cams show room for improvement."

In comparing camshafts it would be easy to be misled by factory timing data. These often appear to be more radical than specially ground cams. The secret of the "less-radical" camshafts that have been reground for additional performance is laid bare by lift vs. crankshaft-movement plots. Remember, too, that duration cannot be considered separately from overlap. Lift vs. crank-degree diagrams will look similar for various cams because the designers must all work within the same limits for valve-train accelerations, spring pressure for returning the valve to the seat, and compatibility of the tappets with the camshaft lobes under the spring loading, lubrication, and temperature conditions which exist. The rate of lift and descent, and therefore the curve shape, will turn out to be decidedly similar on various camshafts, provided the engineers know what they are doing. Be wary of vastly different curve shapes.

When comparing camshafts do not be impressed or swayed by a few degrees difference in timing or several thousandths lift. Such minor differences will not make two similar cams run that much differently.

If you plan to take apart the crankcase to install a 1964 crankshaft or do a complete rebuild, be sure to order your camshaft ahead of time. The grinders turn out hundreds of V8 cams for every Corvair that runs through their factories and a delay of a week

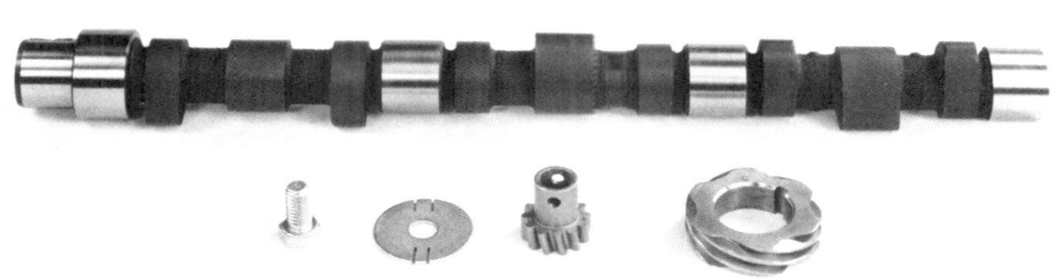

Clark's reverse-rotation kit is available with stock or reground camshafts, left-hand drive gears (one for distributor shaft, other for crank) allow distributor and oil pump to turn in the stock direction even though the crankshaft and camshaft rotations have been reversed. Reversing the engine is the only way that the Corvair can be installed in non-swing-axle VWs or Porsches.

or two or more would not be unusual in getting a camshaft for the Corvair. The Corvair parts specialists, such as Clark's, do not grind cams. But they do have many years of experience about what works well in mildly modified street cars. Isky Cams list several for the Corvair, Clark's has both their own line of cams and the Otto Parts grinds. They will usually have the more common cams in stock. In an all out race engine, talking with the racers is the simplest way to go. They know what works and what doesn't. Most of the cam grinders who have been around for a while have ground Corvair cams in the past, and can do one for you as well. Bill Fisher's friend, Racer Brown retired from the business years ago. Cam grinders today usually regrind old camshaft cores, but would choose to use new parts instead. However, original blanks are almost impossible to find. It would be wise to specify a '64 or later cam core when ordering, even if your present plans do not include the '64 crankshaft installation. The larger throw of the '64-later crank needs more clearance at the lobe and the '64-up cams are special. Be sure to obtain the grinder's recommendations as to spring pressure so you will be able to install the appropriate springs. It is safe to assume that you will be asked to discard the small-diameter 80 or 84 HP springs which will require machining the spring seats in the heads to accommodate the larger springs. Be careful when machining the 140 heads for larger/longer valve springs. On each head one valve spring pocket comes very close to the passageway used by the choke pass through. A tiny opening here will cause an oil leak that will drive you crazy trying to track down the source.

Rockers and Rollers—Past and Present

There were two different designs of stamped steel rockers during the production run of Corvairs. It is only a subtle difference, but it does make a difference for the performance builder. The earlier design used, I believe, until mid-1966, a slightly longer and narrower contact patch for the valve tip. The later design and all of the replacement pieces, uses a wider, shorter contact face. I believe the earlier design better supports the high lift cams used on performance applications. Check out those 80 HP 1960 motors for their used rocker arms. Along with their worn-in rocker ball, they may be the only thing you want to salvage before sending those motors for recycling. The new reproduction rocker arm kits sold by Clark's are a copy of the later GM design. In the 1960s, several manufacturers made roller-trunnion and roller-tip rocker arms for the Corvair. Crane and Norris were the most common suppliers. Both were out of the business by the late '70s. Because of the common use of the Corvair motor in dune buggies and "sand sports" applications, Bob Sutcliffe had the Harlan Sharp company make two different sets of rockers for the Corvair. The 140 HP motor has a slightly different valve angle between the rocker stud and the valve centerline. Harlan Sharp still makes those rockers for the Corvair. They are currently the only supplier. Several people have adapted the Competition Cams ball stud, roller-tip rockers for use on the Corvair. Several items need to be altered for the Corvair application. The Corvair rocker arm, as installed, is upside down from how it is mounted in the V8s. In the V8, oil spits out of the top of the rocker and drains down to the rocker ball for lubrication. In the Corvair, the pushrod is on the bottom, there is no hole in the rocker where the pushrod contacts the rocker. The Corvair rocker has a hole drilled in the side of it. The Corvair pushrod has a hole drilled in the side and, as the pushrod rotates, the oil spits out of the side of the pushrod and squirts up through the hole in the rocker to lubricate the ball. To use the V8 rocker on the Corvair you need to have their V8 oil hole closed off, a new

The tall covers will work with either 4- or 6-hole heads.

The tall cover on the right has plenty of room for roller rockers and stud girdles. It retains the Chevy design for the gasket locating tabs. Check for clearance on exhaust parts.

hole drilled into the side to squirt the oil up and the slot widened to allow the Corvair alignment.

Lifters

Chevrolet Part No. 3799644 was recommended for your stock or reworked engine. This lifter, when preloaded by turning the adjustment nut in ⅛ to ¼ turn from clatter will operate faultlessly beyond 7,000 RPM. Expert Larry Eave points out that hesitation and rough idling, which often accompany turning in the tappet adjustment from clatter, can be avoided by backing off the nut until the clatter occurs, then just eliminating the clatter. Do not establish the preload just then. Go to the next valve and adjust it the same way, and so on through the entire valve mechanism. When all of the valves are at the "just not clattering" point, shut off the engine. Now proceed to establish the ⅛ to ¼ turn preload at each valve adjustment nut. It works! I have also followed that procedure to the point of setting the preload, then letting the motor sit overnight, so the lifters can bleed down the small preload. Although Racer Brown had a good hydraulic tappet for Corvairs, avoid the so-called "fast-leak racing tappets" as these are known to collapse at both high and low RPM. Never run old tappets on a new cam or mix your tappets. Keep each tappet with the cam lobe it has been running on or replace them all.

Different Lifter Brands

Those specific GM hydraulic lifters were replaced in production many years ago. GM supplied several different lifters for use in the Corvair. Some were the deep-dimple design. Several of the same designs are still available but the number of actual manufacturers has dropped to a handful. GM doesn't make any lifters for the Corvair. There are several designs of lifters that will work okay in the Corvair, but nothing is being specifically built for the Corvair. In theory, a Corvair-specific lifter would pass more oil up into the pushrod than the small-block Chevy lifter. In talks with one of the manufacturers, we were told that the original Corvair lifter, as recently tested, did not pass any more oil than the one they supply for the V8s. They make an almost identical lifter for the Harley-Davidson which did pass more oil. To me, it makes sense, in a performance application, to make sure that the lifter design securely retains the inner piston. The only lifters I have seen come apart are the ones with a wire retainer. Then again, they have many millions of miles of good use behind them. If you find ones with inner snap rings, it is good insurance. Belief in one brand or model of lifters is almost a religion, people defend their choices fervently. The equipment needed to produce valve lifters is so complex and of such high precision, that there are no small producers. It is possible that a certain design might provide a "deeper seat" for the pushrod. This might be an assistance in installing pushrods, but if you are twisting the engine to such an RPM that the pushrod could float out, you will have other problems.

Valve Springs and Retainers

Two valve springs served all 1960–63 Corvair requirements. The smaller of the two was used in 80 and 84 HP engines; the larger ones were used in all of the high-performance and Spyder engines. 1964–65 engines used the large spring, which is essentially a "dual" spring because it includes a flat wound damper coil. The damper is an interference fit inside of the round coils to damp out harmonic vibrations. Both stock and special springs are described in the following table. If your heads are of the low HP kind made in '60–'63, you will have to open up the spring seats to 1-¼-inch diameter to accommodate the larger stock or special springs. The smaller springs are absolutely not suitable for high-RPM use. Machining the spring seats to the larger diameter is described in the material that follows on setting valve-spring heights, and also in the "Cylinder Heads" chapter where "all-out" modifications are discussed. Dual springs, long used in racing engines, also provide dampening of harmonics and some special springs afford the extra advantage of material that is better than the stock carbon-steel springs. Special springs listed in the table are made from chrome-vanadium steel. Always follow the recommendations made by the cam grinder who supplied your camshaft, checking to ensure that the springs have adequate coil clearance at full lift as discussed in the following paragraphs. Never take anything for granted!

Valve-spring heights are checked individually at each valve, measuring between the underside of the valve-spring retainer and the top of the shim on which the springs will rest. The valve is installed in the guide and the retainer and valve keepers (keys or locks) are added. Tap the end of the valve stem lightly with a mallet to seat

Rocker arms—Stock, Competition Cams cast roller tip, and Harlan Sharp roller trunnion.

the locks into the retainer's taper bore. Racer Brown makes an excellent suggestion for measuring installed height. Cut a piece of welding rod to the required/desired length and use this to check the distance. Hold the rod parallel to the valve stem. If the rod shows that installed outer-spring height will be greater than specified, add spring washers as shims at the guide to get the dimension corrected. Should height be insufficient, more drastic measures are required. In such instances the spring-seat surface of the head must be lowered by machining with a hollow shell mill with a pilot that will slip fit into the guide bore. Such tools can usually be rented from the cam grinder, or pay the machine shop and have it done correctly. It is important to flood the entire operation with cutting oil as the counterbore is turned by a slow-speed electric drill motor. A drill press is not recommended because of the compound angle of the Corvair guides that complicates holding the head on a drill table. Most of the shims supplied with the valve spring and retainer kits will be approximately 0.020 inch, which means that you can get the installed-spring height accurate within +0.010 inch The installed height should be per the spring maker's specifications and may vary from dimensions that would be used with stock springs. For example, the stock height is 1-$\frac{21}{32}$ inch; Racer Brown says use 1-$\frac{11}{16}$ inch for his damper springs. Be careful about depth of this cut, and check for leaks.

Two other points of possible mechanical interference must be checked for clearance between the spring retainer and the top of the guide at full lift, and clearance between the spring coils at full lift. There should be a minimum of $\frac{1}{16}$ inch between the top of the guide and the lower side of the spring retainer at full lift. This measurement is made without the springs installed. Measuring for coil clearance at full lift is most easily accomplished by carefully holding the spring in a vise, tightening the vise to the recommended installed height, then further tightening it to compress the spring to the equivalent of full lift. A 0.010 inch or thicker feeler gauge should pass through the spring coils at full lift position. Inner springs are checked alone, with the measurements made to the outer-spring seat of the retainer. If the feeler will not pass between the coils, then the installed height must be adjusted to provide clearance between the coils at full lift.

The person who assembles springs onto the valves without checking the height, clearance between the coils and at the guide in the full-open position is just asking for a damaged engine. You cannot assume that "everything will average out."

One of the simplest additions that you can make for increased RPM is changing the stock stamped-steel valve-spring retainers for lightweight retainers. The stock retainers weigh 23 grams; special ones are 9 grams or less. Aluminum alloy or titanium ones are available. While it is possible to change these without removing the cylinder heads, it is better to remove the heads

C-clamp valve spring compressor is needed for assembly and disassembly of valve springs.

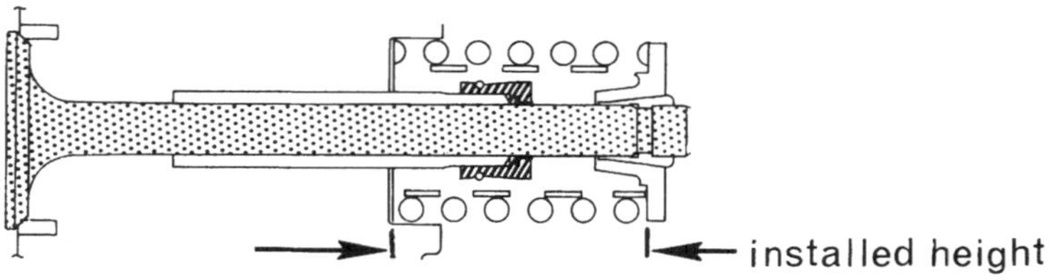

VALVE-SPRING SPECIFICATIONS

Type Spring	*Free Length*	*Seat Pressure*	*Pressure at Lift*
Chevrolet 1960–63, 80 and 84 HP	1.74"	60 lbs.	145 lbs. @ 0.344"
Chevrolet 1960–63 Hi-Perf. and Spyder and all 1964–69 models	2.08"	82 lbs.	175 lbs. @ 0.391"
Racer Brown, 1-¼ damper type	2.0"	95 lbs. @ 1.6875"	240 lbs. @ 0.437"
Isky PN #205-D	–	115 lbs. @ 1.70"	268 lbs. @ 0.490"

to forego any possibility of bending the valve stems during installation. The angled Corvair valves make use of a "C-clamp" valve-spring compressor absolutely essential. Special retainers for Chevy V8s work with the larger Corvair springs.

Have no fear about using aluminum or even titanium in this highly stressed area of your engine. Such retainers have been used for many years in motorcycles, but it took auto-engine modifiers a while to realize that what "couldn't be done" had been working successfully all along. Both aluminum and titanium retainers are available from Corvair parts sources. There are also some lightweight steel retainers designed for the small block Chevy application which will work here.

Stock valve keys (keepers) are more than adequate for any spring pressures that will be encountered in the Corvair, regardless of cam grind, HP, or RPM. Forward Controls utilized a different design of valve keepers on most motors.

Cams with Roller Lifters

Corvair's crankcase design does not lend itself to simple installation of a roller-tappet camshaft. There is barely enough space for the guiding devices required by most roller tappets. Also, the usual roller tappet has a large notch in which the needle-bearing roller is installed. This gap could create a mammoth oil leak from the main oil galleries that supply both the hydraulic tappets and the crankshaft. For this reason, no roller tappet cams are normally cataloged for Corvair. Some Corvair owners who insist on using roller-tappet cams are rerouting the crankshaft oil supply and using the roller tappets that self-guide against one edge of the cam lobe. New applications are always happening. There are several roller-cam Corvair motors running today. They are being tested and evaluated to see if the work involved in adapting these lifters, and the much more aggressive cam profiles that can be used, provide enough of a benefit to justify the cost.

A great amount of space has been devoted to a description of installing bigger valves. You probably wondered as you read these details in the "Cylinder Heads" chapter if it is worth the time and money. It most certainly is, especially when you are installing a special camshaft. The larger valves help you get your "money's worth" in performance. The combination of big valves and a special camshaft should add 40 to 50 HP to any Corvair engine if properly installed and tuned. That kind of horsepower cannot be overlooked by any serious competitor.

The gain in power afforded by larger valves is attributable to the added mixture-flow area (circumference × lift) when the valve is opened. With a larger-than-stock valve, the area is greater at every portion of the lift curve. The drawing here shows a comparison between the available area at full lift when comparing the stock 1.35-inch intake with 1.5 and 1.7-inch OD valves. Intake opening at various lifts has also been plotted for the three valves, using 1.5:1 rocker arms and a Racer Brown Super Street H-9 Camshaft. The opening has been taken at the center of a 0.060-inch-wide seat in every instance. Bigger exhaust valves are equally important in making these engines breathe, especially at high RPM. Regardless of the type of cam you choose, install larger valves if the budget will possibly permit you to do so.

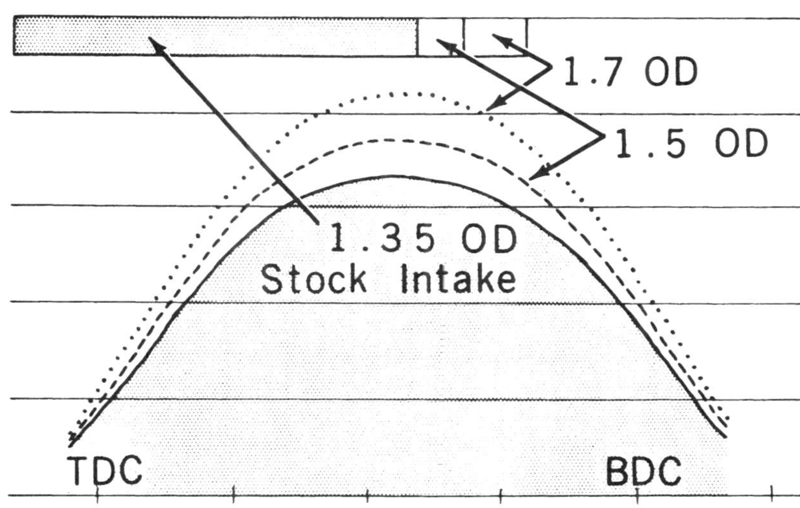

1965–67 140 HP engine valves dwarf other Corvair valves.

8

Pistons, Pins, and Rings

The factory Corvair pistons are conventional flathead aluminum alloy with cast-in steel struts. Each slotted-oil-groove slipper-skirted autothermic piston is lightly tin plated to reduce break-in scuffing. Room-temperature clearance measured at skirt center is only 0.0011 to 0.0015 inch. Three piston designs—all permanent-molded—have been used as detailed in the Stock Piston Comparison table. The 1964–69 pistons employ a revised ring-groove arrangement (thinner rings) and shorter compression height to compensate for the longer stroke.

Because Corvair's 1964 increase in swept volume to 164 inches was obtained with the longest-stroke crank that conveniently fits the crankcase dimensions, the most common route for adding cubic inches is through a larger bore. Meanwhile, GM discouraged larger bore sizes, and simultaneously improved their own spare-parts sales by two means: (1) not listing pistons either in standard bores or oversizes, except as complete assemblies, and (2) bargain pricing these stock-bore replacement assemblies. Each assembly, consisting of a cylinder complete with piston, pin, and rings, originally cost a mere $15.50. Low-cost rebuilds with these items strongly appealed to Mr. Average Man and Miss Old-Maid Schoolmarm with worn-out Corvairs. But to the hotrodder who learned long ago that cubic inches are the only viceless hop-up trick, a bigger bore is the logical answer. "When in doubt, bore it out!" With the Corvair's separate "captured cylinder" design, the limit to the overbore is eventually the hole pattern of the clamping studs. An overbore of the original cylinder results in less clamping area for the head gasket. Thinner gaskets have been available to keep from hanging part of the gasket into the combustion chamber, an all-around bad thing.

Remember, of course, that adding displacement increases the compression ratio. Complete details on figuring compression ratios are found in the "Cylinder Heads" chapter.

Available stock, replacement, and racing pistons, pins, and rings are discussed in this chapter. Clearances, design techniques, and installation tips that you need to know are covered thoroughly.

Stock Pistons

Although the stock pistons are described in the summary, the table on the next page shows the relationship of the various bore sizes, compression heights, and ring arrangements.

Graphic proof of changes in stock pistons. '64 and later style piston at left has reduced compression height and cutaway under pin boss to compensate for lengthened stroke. Narrow rings work better at high RPM. Piston on right is '61-'63 type.

Chevrolet's own fine pistons are used to advantage when displacement is stock-size limited or when racing regulations stipulate stock parts. Even then, if you plan to use the engine for any high speeds, more clearance should

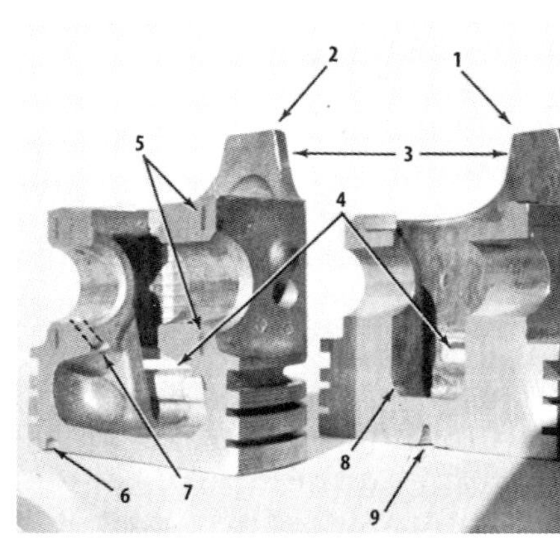

Sectioned pistons show construction features: 1. Forged type. 2. Corvair stock model. 3. Slipper-skirt design. 4. Slotted oil grooves. 5. Steel anti-expansion strut. 6. Locater notch. 7. Oil hole to pin (both pin bosses in each piston). 8. Pin support. 9. Grinding center.

Another comparison photo to show the differences between the two compression heights used on stock Corvair pistons. Note obvious cast-in relief for rod-cap clearance (arrow).

STOCK PISTON COMPARISON: All dimensions are in inches

Year	Displace-ment	Bore	Compression Height	Compression Ring Width	Oil-Control Ring Details
1960	140	3.375	1.761	0.077	0.186 1-pc. cast iron
1961	145	3.437	1.761	0.077	0.186 1-pc. cast iron
1962	145	3.437	1.761	0.077	0.186 3-pc. RC_1
1963	145	3.437	1.761	0.077 TCS_2	0.186 3-pc. RC_1
1964	164	3.437	1.591	0.062 TC_3	0.125 3-pc. RC_1
1965–1969	164	3.437	1.591	0.062 TC_4	0.125 3-pc. RC_1
1966–up	164 with A.I.R.	3.437	Note	0.062 TC_4	0.125 3-pc. RC_1

NOTES: RC_1 = chrome-plated rails, all models of that year.
TCS_2 = top compression ring chromed on Spyders only.
TCS_3 = top compression ring chromed on all models.
TCS_4 = chromed top compression ring on 95- and 110 HP models; high-strength ductile iron, molybdenum-coated used for top ring on 140- and 180 HP models.
Skirt length: 1960–63 = 3.358; 1964–69 = 3.190.
Width between outside edges of flat areas for rod clearance: 1960–63 = 2.705; 1964–69 = 2.655, plus ¼-inch radius under pin as depicted in accompanying photographs. Many engines with Smog Pumps (A.I.R.) used pop-up pistons to regain compression when "open" style combustion chambers were adopted for emission-reduction reasons.

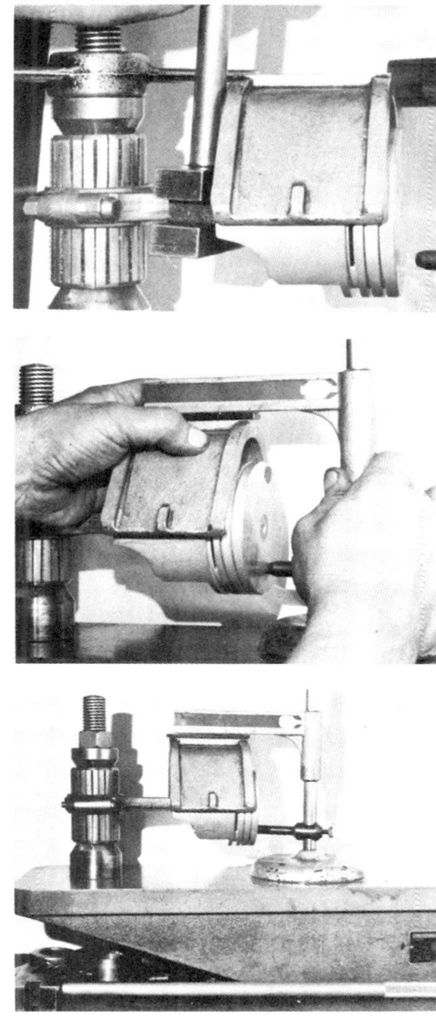

Checking piston and rod alignment with K.O. Lee fixture. This is an essential step in precision engine assembly. Top photo shows bending or twisting rod with special tool. Some brand new rods are both bent and twisted.

be provided. Remove the heads and cylinders, and add clearance by honing. This can be done without removing the pistons from the crankshaft, but like the old Indian scouts, look for signs. Check the wear pattern on each piston to make sure that it is parallel with the piston-skirt centerline. If not, then this is your sign that the piston is cocked and not traveling straight in the cylinder. Any piston showing uneven wear must be removed for alignment with its connecting rod. Twisted or cocked piston/rod assemblies steal horsepower through unnecessary friction and create unwanted heat and wear.

If you really crave peak performance, remove all of the piston/rod assemblies and check for alignment with tools available for that purpose at automotive machine shops. This will provide you with an opportunity to check bearing wear, etc. Follow shop-manual instructions to check ring grooves for wear, cleaning them, and check both ring-gap and ring-groove clearances.

Replacement Pistons and Racing Pistons

Some replacement pistons can be used for racing, but not all racing pistons are suitable replacements for the stock items. Suitability of the various types depends on the engine's intended use. The original Corvair piston was subject to failure in hard use. The piston top would collapse if encountering detonation and the piston top would come off, breaking loose at the lower ring groove. Many years later, several types of replacement pistons are now available for the Corvair, including sand-cast, permanent-molded, hypereutectic cast, and forged types. Forged pistons are made by forcing aluminum slugs into a piston form under extremely high pressures to create a very dense grain structure in the finished part. Forged pistons are claimed to have up to 70% more strength and better heat dissipation characteristics than conventional sand-cast pistons. Their strength at

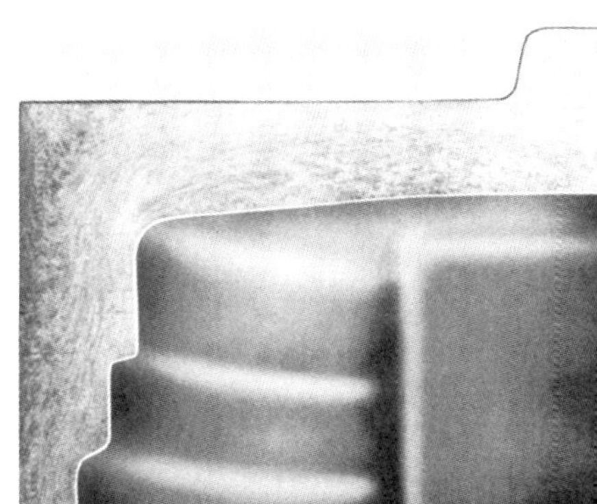

Metal grain flow indicated here show shows how forging adds strength to a piston. Smooth inside surface of piston is created by high-pressure forming with precision tooling. Photo courtesy TRW-Thompson Products.

temperatures over 600° F is superior to that of any other type of piston. Yenko Sportscars supplied Forgedtrue brand of forged pistons in their more powerful Stinger motors. Those forged pistons used the wider rings as supplied for the early model Corvairs.

Running a normally aspirated (non-supercharged) engine on gasoline with careful control of mixture strength should permit the use of almost any currently available piston for the Corvair. However, gasoline burning supercharged (or turbocharged) engines or non-supercharged engines using nitrated fuels, should be equipped with forged pistons to take advantage of the extra strength which these pistons provide. For instance, the Spyder engine has an overheat-warning switch that indicates when the head temperature has reached 575° F. Head-temperature indication is probably a lot lower than that being suffered by the piston crowns. You can speculate that the piston temperature is probably approaching 650° F to 700° F by the time the warning switch tells you about the problem. When you consider that pistons are forged at only 900° F, you can see that the piston crown is starting to soften at the temperatures which are often obtained in all-out competition, especially in the air-cooled Corvair.

Any replacement piston must be designed with the high temperature Corvair combustion chamber in mind. The compression height must be matched to the stroke of the motor, if a late piston is used on the short throw crank, you will have the lowest compression in town. This wrong compression height would also give excessive deck clearance, thereby reducing performance by eliminating the essential squish turbulence from the combustion chamber. Additional data on turbulence and its relation to deck clearance are in the "Cylinder Heads" chapter. These data are extremely important to tuners seeking peak performance.

Some manufacturers modify round pistons to permit their use in Corvairs. Round pistons are okay if the area under the pin bosses is properly machined. Check the comparison table and stock-piston photos and measure the ones that you plan to install before assembling them onto your rods.

If you are planning a larger-than-stock engine that will be primarily used for transportation with only occasional hard use, then the forged replacement with slotted oil groove will probably be your first choice. These pistons offer an excellent compromise between strength, quiet running, and cost. TRW, now marketed by Speed Pro, makes these pistons in stock and oversize bores at reasonable prices. The TRW/SpeedPro L-2206F is available to fit 1964–69 engines or earlier engines being fitted with the '64 or later crankshaft, in sizes up to .060 oversize. Because they are available, buy the largest. Take advantage of the situation by utilizing as much of this cheap horsepower as you possibly can. I have noticed the quality of these TRW/SpeedPro pistons has dropped off as the production quantities have shrunken.

If you are planning on building a motor for performance, there is no reason not to build the largest displacement Corvair motor you can, consistent with your budget and planned application. The 1964-up motor, with its added stroke, has no downside to it. With the correct parts, the 164-inch motor can be fitted into any model Corvair, even the earliest. As long as mild compression is maintained, there is no sacrifice in longevity or fuel requirements.

The TRW Part No. L2156F was used as a replacement in 1960-63, small displacement Corvair engines. Their "semi" pistons can be finished to fit a 3.500-inch bore, but check closely on ring availability before buying the pistons or having a semi set made to your specifications. That TRW piston part number may no longer be available. All the regularly available pistons for the short-stroke motor are now cast pistons. Recently, several new pistons have become available from the largest Corvair parts suppliers. New materials and construction techniques will assure that the Corvair will receive the benefit of modern technology. Features such as pre-machined valve clearance eyebrows and drilled oil return holes are nice extras, but are not needed for regular street use. Each type of piston material has its good features. Although forged pistons are ultimately stronger than cast pistons, the mechanical design of the forged piston causes them to grow in

The new forged pistons from Clark's Corvair Parts use drilled oil return holes beneath the oil control rings. This is a weak point in most Corvair pistons, where a milled slot removes some of the support for the piston head on both sides of the piston.

size as they heat up—more than a properly designed hypereutectic cast piston. These new hypereutectic pistons are cast with a high amount of silicon mixed into the aluminum material. Because these pistons do not "grow" as much in size as they heat up, you are able to use tighter piston to cylinder wall clearance. This results in quieter operation (less piston slap) and better longevity for piston rings. Clark's Corvair has also brought a new forged piston to market. It uses drilled oil return holes instead of the slots of the regular pistons, always a vulnerable point of failure.

In addition, the wide-spread availability of hop-up parts for the air-cooled VW motor has resulted in an increase in the cross-over parts, parts which can be used as-is, or slightly modified for use in the Corvair motor. The Mahle piston, German designed, is now available with a modified replacement cylinder, designed as a Corvair drop-in kit. If you get much larger than stock cylinder outside diameters, machine work on the cases and the heads will be needed.

Real Racing Pistons

I talked recently with the owner of one of the big-name racing piston suppliers. He told me about a revolution in the industry of the last 10 years. Pistons used to be one of the flagship products of high-speed, large volume manufacturing. The process of forging aluminum blanks and machining hundreds of identical pistons took large machine planning and industrial giants. The new CNC (Computer Numeric Control) machining capabilities of even smaller production facilities have made short-run, custom pistons a "reasonable" possibility. Buying only six of one style will remain an expensive option, but production runs of a couple of dozen units have become possible. Several times racers have grouped together and bought a few sets at a time. The piston manufacturers have redesigned their production facilities to better serve the smaller quantity, quick turnaround needs of racers. The prices are still high, compared to the high-volume units, but the quality and the capability of designing for special piston-top layouts or ring-types is there. These are not needed for most street use, but the racers have more specialized needs.

Skirt Knurling and Grooving

Numerous articles have appeared illustrating engines being assembled with new pistons that were knurled to decrease clearance. *Knurling*, as usually practiced in rebuilding engines, deforms the skirt to decrease side clearance and reduce noise from piston slap. Because of the deformation that occurs, new pistons should not be knurled on the skirts.

Shallow grooves on the piston skirt can be added instead of knurls to keep oil on the sliding surfaces. Forgedtrue pistons used on higher stage Yenko Stinger motors used pistons with these grooves.

Piston Balancing

Any replacement piston for the Corvair will probably weigh more than the stock item it replaces. This is not a problem because the Corvair's cylinder arrangement provides "inherent balancing," just like an inline six-cylinder engine. All piston and pin assemblies must weigh the same, of course. They should be kept as close to the stock weight of 453 grams (1 lb = 454 grams) as possible without reducing strength in vital areas. Avoiding excessive piston weight improves engine life by reducing loads on the piston itself, pin, connecting rod, bearings, and the connecting-rod bolts. It is important to have balancing done by a capable, experienced organization. Material should not be removed from the slipper portion of the skirts or the pistons will be seriously weakened. Balancing pads are provided on most pistons for this purpose.

PISTON CLEARANCES

	Piston Design Type			
	Stock Corvair	TRW/Speed Pro	KB Hypereutectic	Clark's Forged
Road	0.0015–0.003	0.005	0.0015–0.003	0.0015–0.003
Dual Usage	0.003–0.004	0.005	0.002	0.0015–0.003
Racing	0.003–0.004	0.006–0.007	0.0045 max	0.0015–0.003

Clearances

Clearances are varied according to the materials and manufacturing specifications of the piston maker, as well as the type of use to be given the engine. In the case of stock pistons, these are fitted quite closely because two steel anti-expansion struts are molded into the piston. Slotted-oil-groove pistons without such struts are also fit closely. Mackerle's book points out, "By slotting the piston below the scraper ring, the transfer of heat from the piston crown to piston skirt is prevented and the influence of piston crown temperature on skirt distortion is thereby limited, the return flow of oil from the scraper ring also being improved."

When ordering pistons for your Corvair, specify the type of use to be given the engine, and indicate the desired cylinder-bore size. The piston maker will provide appropriate clearance for that bore size and application. Extremely high compression motors will require custom piston crowns.

If you use the replacement-type pistons then these will have the clearance for normal-driving use with the specified bore size. For applications requiring greater clearances, have the pistons available when the machinist is boring the cylinders. Specify the desired clearance. He will increase the nominal bore size, usually by honing, to provide the added clearance. Boring is usually to a size just under the finished size with honing used to get the final dimension.

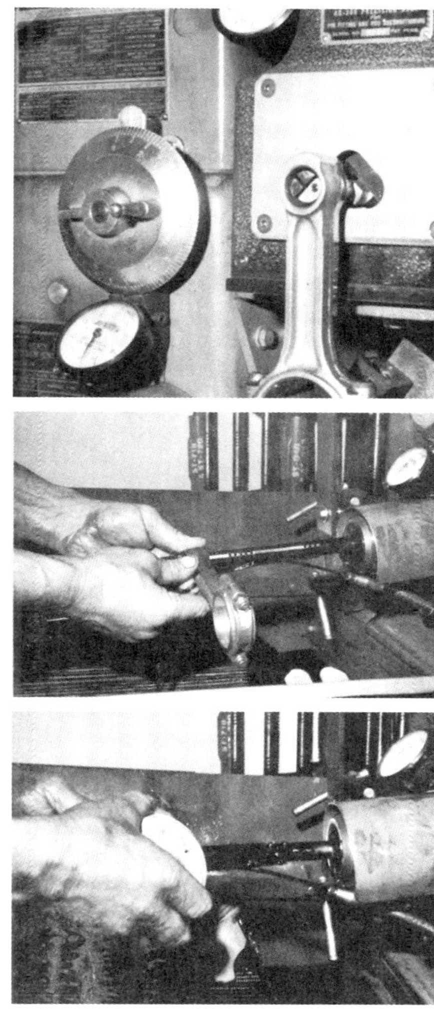

A Sunnen hone equipped shop is best suited to handle the job of Corvair pin fitting to the rods and pistons. Top photo is of rod measurement to determine how much of it must be honed out for proper interference fit. Center photo shows honing new 1964 rod eye. Bottom photo: pistons, even pin-fitted ones, sometimes require honing for additional clearance.

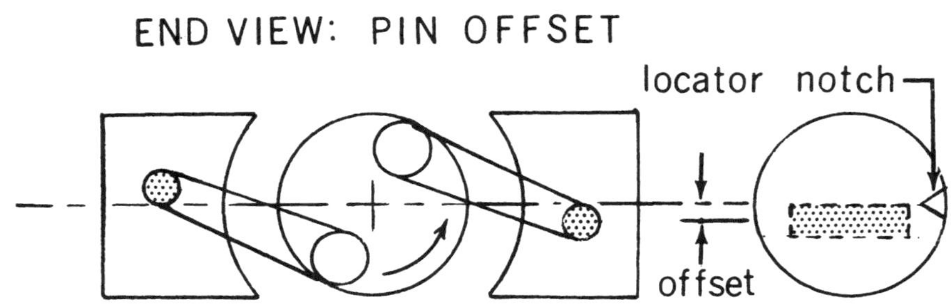

Pin is offset within the piston against the direction of rotation (see arrow).

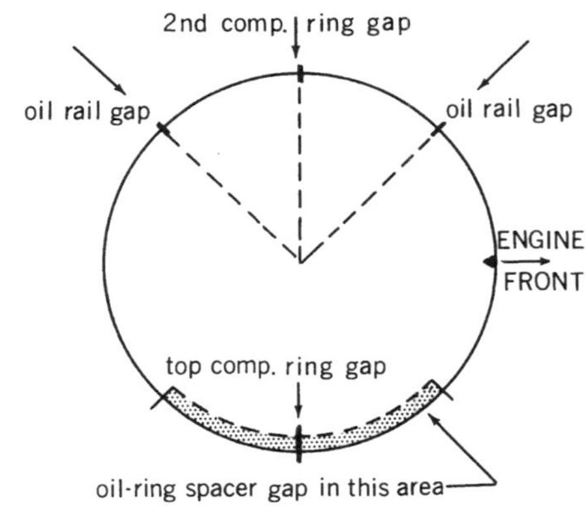

Staggering compression and oil ring gaps on all pistons gives maximum oil economy. It also places rings to clear the cylinder notch at the base of 1964-69 cylinders. Drawing copied from a dealer service bulletin.

Solid-skirt pistons of either forged or sand-cast types require greater clearances. Remember that as clearances are increased, oil control becomes somewhat more difficult, ring wear increases due to piston rocking, and the engine becomes noisier. Be sure to follow the clearance recommendations of the piston manufacturer. The Piston Clearances table is a guide and indicates clearances as currently recommended by the piston firms mentioned.

Pressing pin into piston and rod requires special fixtures to ensure that pin enters rod and piston bores correctly. Chamfer inner edge of pin bore in piston to assist pin entrance into bore as it comes through the rod eye.

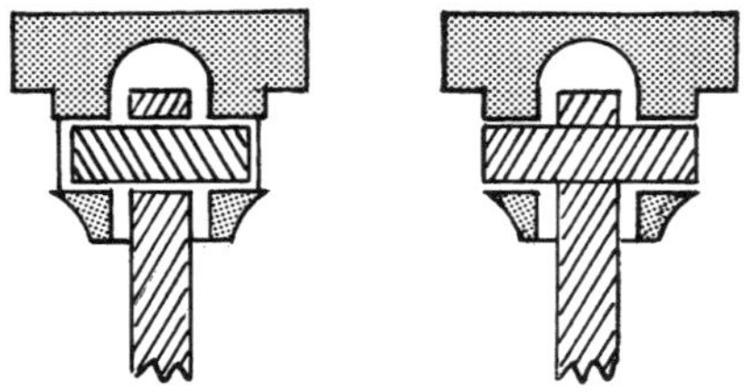

Full-floating pin (left) is not desired because there is no rigid structure to resist piston cocking. Loads are taken at only one rod surface. Rapid wear and noise result. Pressed-in stock-pin arrangement at right requires careful installation but results in rigid "T" structure with piston loads taken at two bearing surfaces on pin.

Piston-Pin Offset and Fitting

Stock Corvair piston-pin bosses are offset from the piston centerline by 0.055 to 0.065 inch. Mackerle says, "A piston pin offset by about 1.5 mm (0.059 inch) against the direction of rotation increases the period during which the load on the piston is transferred from one side to the other, thus reducing piston slap at top dead center." If the engine is being reverse-rotated for engine swap purposes, the pin should be offset the other direction, the piston turned around.

It is easy to determine how the stock pistons are to be installed to take advantage of this offset: the notch in the top of the piston always faces forward toward the flywheel end of the engine. It is not quite so simple on pistons without such identifying marks. You can be sure you are installing the pistons correctly if you: (1) measure the pin bores and mark the side of the piston closest to the pin; (2) measure the pin bores and mark the piston with an arrow or forward mark to simulate that used on stock pistons; or (3) install the pistons with the pin closest to the bottom side of cylinders 1, 3, 5, and closest to the top of cylinders 2, 4, 6. Note illustrative sketch.

Replacement-type pistons are usually pin fitted with extreme accuracy by the manufacturer. For this reason, it is essential to keep the individual pins with the pistons into which the maker fitted them at all times—they are selectively fitted and are not interchangeable.

It can be necessary to hone the rod "eyes" to reduce the interference fit of the pins to an acceptable level. Too much interference will gall or tear the rod and pin surface. This causes rapid pin-bore wear in the piston. The press fit should be in the range of 0.0008 to 0.0012 inch. All pins are not the same diameter. Some pins require that the rods be honed for proper fit. TRW recommends that oscillating pins (stock arrangement) be given 0.0003 to 0.0006 clearance in the piston. Full-floating pins require 0.0001 to 0.0003 inch clearance in the piston. More pistons are ruined by faulty assembly techniques than by any other cause. It is extremely important to ensure that the machinist uses the correct fixtures, measures pins and rods, and follows this recommended procedure after establishing the correct fits. If heating of the rod is recommended by the piston manufacturer, abide by their request. Use the proper fixture to automatically center the pin and rod to the piston and act as a pin guide. Lubricate pin and rod eye. Support piston squarely and firmly. Enter pin into connecting rod with the hydraulic press. Release pressure. Recheck pin/rod alignment by eye. Apply pressure again if all is going well. The idea is to make sure that the pin does not broach out the pin bore in the rod, thereby galling the pin and rod. Press the pin through to the installed position. The rod should be centered on the pin. Check pin and rod assembly alignment with appropriate fixtures. This is essential to make sure that the piston will be traveling straight in the assembled cylinder.

WARNING: *If proper fixtures are not available, do not attempt to install pistons onto rods. Rods and/or pistons can be destroyed easily when wrong procedures and tools are used.*

Some Corvair articles tell you to use full floating piston pins. This means that the piston pin has clearance in the piston and in the connecting rod, as opposed to the stock setup that presses the piston pin into the rod. Some builders bore out the small end of the connecting rod and fit a bronze bushing. Others hone the small end for pin clearance, add oil holes, and allow the rod end to oscillate directly on the piston pin's hardened surface. Because the pin is no longer held in the rod, keeper grooves are added to the pin bosses and keepers retain the pin.

This assembly method should only be considered for an all-out racing engine that is to be overhauled regularly and often. There are some advantages, of course. First, the engine is easier to assemble because special tools are not required to install the piston onto the connecting rod. Secondly, the floating pins probably contribute a minute reduction in operating friction. Finally, such installations protect the piston from damage during assembly by "hammer mechanics." However, the use of full floating pins, as opposed to the stock arrangement, reduces pin, piston, and connecting-rod life. Because it is less rigid from every standpoint, the assembly is prone to wobble at its three clearance points. And pin locks have been known to come out, or be left out by mistake during assembly, with consequent damage.

The "T" assembly provided by the stock pin pressed into the rod is the better solution because it resists piston cocking, runs quieter, and reduces wear. I would recommend that you should use it, even in all-out racing engines.

Piston Rings

Piston rings are always a subject of interest among auto enthusiasts. This is because they are almost always replaced during engine rebuilds and must, of course, be replaced if the bore size is changed. Although few mechanics hesitate to debate the merits of various brands of piston rings, many do not understand the basic functioning of these vitally essential parts.

The sealing of combustion and compression pressures occurs as these same pressures enter the ring grooves forcing the rings down against the bottom of the grooves and outward against the cylinders. Although outward stresses are built into the rings, these merely assist the gasses in sealing. Ring pressure against the wall does not accomplish much sealing. Tests made by Ricardo showed that rings operating without gas pressure behind them failed in a few minutes of operation.

Ring-groove condition is vital to the functioning of the rings. According to Richard MacCoon of Grant Rings, "Rings are just like valves and must have a good seat on which to seal. Groove finish is every bit as important as that of the cylinder or the ring itself. Piston makers machine piston-ring grooves with extreme care to ensure perfectly smooth surfaces. Yet you will often see 'mechanics' use a broken ring to clean out carbon from these finely machined grooves. They gouge and burr the groove and still expect the rings to work. Special tools for reconditioning grooves should always be used when fitting new rings to used pistons."

As shown in the piston table, many Corvairs are equipped with chrome plated top-compression rings and oil-ring rails. The factory can use chrome rings because they are able to tightly control cylinder-wall finish. Thus, they get fast seating and can take advantage of the corrosion resistance and cylinder-wear improvement afforded by chrome rings. That close control of machining is the reason that new cars can go 200,000 miles or more.

Chromed rings, new or used, must always be examined carefully for chipped or damaged surfaces prior to their use/reuse. Such defects cause serious cylinder-wall gouging if run in an engine. Even new factory fresh chrome rings are sometimes chipped. When reringing an engine with any wear on the cylinders it is important that you use non-chromed cast-iron rings because the worn cylinders are no longer round. Chrome rings cannot provide perfect sealing in worn cylinders because their hard surfaces do not readily conform by "wearing in."

Racing-engine builders frequently talk of "ring flutter." At high RPMs, the top ring may have sufficient inertia as

This style of ring compressor allows a cylinder to be reinstalled on a piston that is in place on the motor. Handy tool.

Use the piston to insert rings squarely into cylinder bore for ring-gap checking.

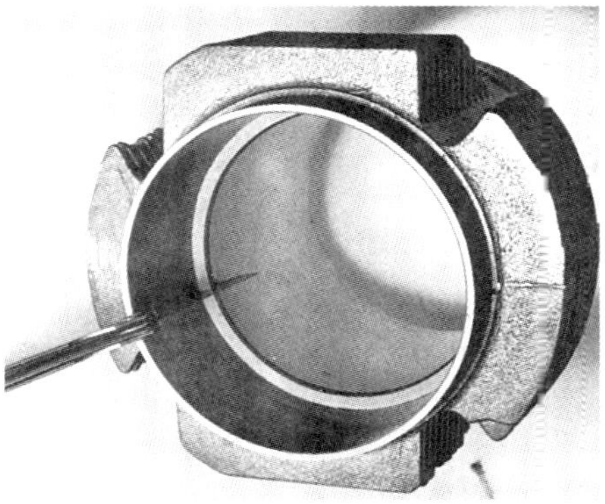

Feeler gauge is used to check gap of each ring before installing rings onto pistons.

the piston decelerates near TDC, that it is held against the top of the ring groove. Gas behind the ring escapes to the crankcase and the top ring collapses inward under cylinder pressure, instead of being pushed outward as in normal operation. This results in vastly increased blow-by which pushes oil out of the crankcase, and drastically reduces power output. Ring collapse or "flutter" occurs only on the power stroke and will cause ring breakage if allowed to continue.

Designers of modern high-performance, high-RPM engines avoid ring flutter by reducing compression-ring width, thereby decreasing their weight and consequent inertia. Ricardo's famous book says, "Taking all of the arguments into account it would seem that the width of the ring should depend on the normal speed of the engine; the higher the speed, the narrower the ring." As proof of this comment we see high-speed two-cycle engines, such as racing outboard and go-kart engines, using 1⁄32-inch-wide rings to prevent flutter at speeds that may exceed 10,000 RPM.

A trend began to grow very fast in the sixties: racing-engine builders turned to a 20-year-old idea to solve their ring-flutter problems on big bore high-horsepower engines. They began installing an L-Shaped ring invented by Dr. Paul de K. Dykes of Cambridge University. Even so, because the rings use low tangential pressure against the cylinder wall, cylinder wear may be reduced during those strokes when the rings are not pressure-loaded.

Another ring manufacturer produces a full line of rings for Corvair models. Total Seal produces a ring set that uses a special interlocking ring design to eliminate an open ring gap on either the upper or lower compression rings. These rings have been used on several racing Stingers with fine results.

Oil control is a separate subject. Dr. Dykes has also been responsible for research that proved that oil passes the rings in both directions and in large quantities. Although oil flows past the oil-control rings on the downstroke, it is returned to the sump on the upstroke. There are at least two types of oil-control rings to consider for use in the Corvair: (1) the stock type using two chrome-plated rails above and below a wavy ring separator, which probably should not be used for rering jobs, and (2) the dovetail design slotted ring with an expander, typified by the construction of the Grant oil ring.

Fitting Rings

In addition to the usual dimensioning of the rings to fit bore diameter and ring-groove width, two other measurements are quite important: ring end gap, and groove clearance.

End gap is measured by placing the ring in the cylinder, using a piston to push it in about two inches from the bottom of the cylinder. This is the area of least wear. If done at top and with proper gap, ring ends may butt together at bottom. This is to make the ring square with the bore. A feeler-gauge set can be used to measure the end gap. End gap should be not less than 0.0045 inch per inch of bore. Example: for a 3-9/16 bore: $0.0045 \times 3.562 = 0.016$. It is better to err on the high side and get too much end gap than too little. A gap of 0.016 to 0.020 is suitable for this bore size, and a gap of .01732 inch will not affect performance. In a pinch, the end gap can be increased by holding the ring in a vise and carefully filing the ends. There are special tools available for truly machining the end gap. The tool seen here allows you to accurately file the ring gap. (Summit Part No: 906795). The center wheel grinds away both sides of the gap as the ring sits on the flat surface.

Ring-groove clearance is also measured easily with a feeler-gauge set. Stock-piston-groove clearance for compression rings should be from 0.0012 to 0.0032 inch, measured between the ring and land. Oil-ring clearance, measured between ring and land should be 0.002 to 0.0035 inch. Replacement and racing pistons often specify 0.002- to 0.004-inch clearance for compression rings. Special groove clearances are needed for most Dykes-type rings, talk with the ring and piston manufacturer if you have a special application. Ring-groove clearances on the low side improve oil control and high-speed performance, but increase the risk of ring sticking in a short period of time, especially where the engine is used only on the street. Obviously, too much groove (side) clearance increases oil consumption, causes groove pounding, and finally ring breakage. Additional side clearance can be obtained when needed by sanding the side of the ring on a flat surface.

Manley Performance Products has a great write-up on locating rings in specific positions. They supply rings for the horizontally opposed Subaru motor with special pins to fix the ring gap at the top. The intent is to better control oil at the bottom of the cylinder.

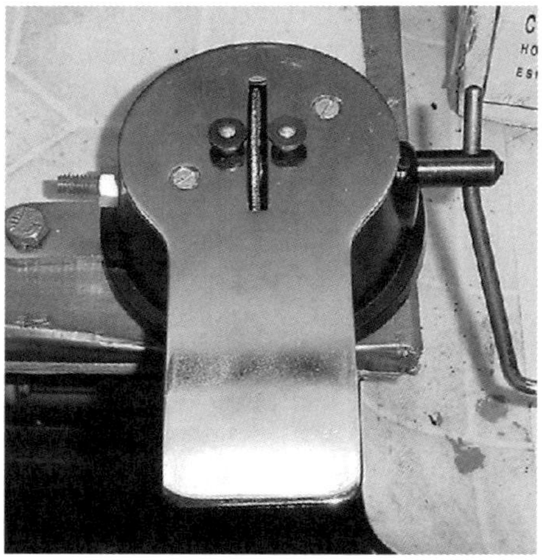

Summit racing sells this tool for opening up the ring gap during fitment. The center wheel carrier is clamped in a vise and the rings sit on the platen, on either side of the cutter.

9

Cylinders and Boring

Boring six Corvair cylinders usually costs more than boring an inline six-cylinder engine. Why? Because each cylinder must be handled individually, with a complete setup for each one. Many shops cannot correctly bore Corvair cylinders because they have not made the necessary modifications to their boring-bar base plates. Although quite similar to motorcycle cylinders, Corvair cylinders cannot always be bored by motorcycle shops because their equipment is usually for smaller bores. When I first had Corvair cylinders bored at the local auto machine shop, they had to make special holding fixtures for their base plate. They only charged $6 per cylinder including honing back in the '60s. Many machine shops have built their own fixtures, because of the part similarity to the VW cylinder. Engine shops around the country that do a lot of Corvair work have usually established a relationship with a local shop in their area to do cylinder boring. Look for a Sunnen CK-10 or newer power hone. This is one area where practice make perfect and an easy method is to buy a set of cylinders from a Corvair vendor when you buy your pistons. Usually a core charge will apply, which is refunded after you return your old cylinders.

Another alternative is to obtain old cylinders that would otherwise be destroyed or junked. This will allow you to bore your cylinders before disassembling your engine, thereby reducing rebuilding time. Make sure there is enough "meat" left to allow you to bore to your desired size and inspect the top head gasket-sealing surface for signs of damage. A single broken off cooling fin

Top and side views of a 1964 Corvair stock cylinder (left) and a 3-9/16-inch bore cylinder (right). Notch in lower edge of cylinder matches relief in crankcase to provide connecting rod clearance. 1965–69 cylinders have a wider head-gasket sealing surface.

tip is nothing to worry about. Assure yourself that the correct-year cylinders are being supplied. The 1960 is unique, with full fins. The 1961–63 cylinders do not have the cutaway clearance at the bottom to allow for the longer stroke. The 1964 and the 1965–69 look similar, but to tell a 1964 from a 1965–69, you must either measure the outside diameter at the top of the cylinder or perform a trial fit into the top of the year cylinder head you will be using. It should be snug. Chevy actually supplied modified '64 cylinders with a spacer ring around the top as replacements for later cylinders.

These are the new full-finned cylinders from Clark's Corvair Parts. Having control of production allows proper fitments for all years and models.

If using a local machine shop for oversize bores, when you go to the shop bring along the pistons you intend to use and at least two extra cylinders. The pistons will be used by the machinist to get the exact clearances you specify and the extra cylinders will allow him to finish the job should one or more cylinders prove faulty. Expect to pay for boring any cylinder that proves faulty.

3-9⁄16 inches is the maximum bore size you should consider for stock Corvair cylinders. Even at this size some cylinders may "hole thru" when bored. A few minor pits in the cylinder surfaces are acceptable, provided they do not go clear through the cylinder wall. A more-conservative 3-1⁄2-inch bore would be less subject to distortion and could conceivably provide longer engine life for a road machine. The special cylinders made by Salih fit into the original head and block openings and were distortion-free at the 3-9⁄16-inch bore size, but are no longer available new. There have been several other suppliers of aluminum cast over steel tube cylinders in the past. If only duplicating the stock size, the advantages would be in reduced weight and better heat dissipation.

There is a practical limit on oversize bores. Anything larger than 3-9⁄16 inches gets you into the realm of expensively machined custom-made cylinders and pistons. The expense does not stop there, the gasket register bores in the heads and the cylinder openings in the crankcase must be milled out to new dimensions. Crankcase openings should not be increased beyond 3.900 to keep adequate seating surface for the cylinder. Such conversions also require special gaskets for the cylinder bases and heads. Carefully consider the total cost of such a project before beginning, including the need for additional oil cooling if the car is to be used for transportation or road racing. The Corvair cooling system of air blown over the fins on the heads and cylinders can remove only so much heat. If your heat input exceeds the cooling system, you are on a sure road to engine problems.

The growth in the last 30 years in the air-cooled Volkswagen hot rod industry has spawned many aftermarket pieces that can readily adapt to the Corvair. There are at least two suppliers of kits that increase the displacement of the Corvair engine to as much as 3.2 liters. The biggest displacement kits come complete with aftermarket VW cylinders and modified rods, and require machining of several points on the heads and cases. Because of these machining modifications, these kits are usually sold on an exchange basis, or you may be able to supply yours and wait!

There are other sources of brand new cylinders. Clark's Corvair Parts is now delivering a brand new iron cylinder. It mimics the 1960 fully finned cylinder design to provide better cooling. Clark's supplies them to fit into either early model or late model cases and heads and they are offering them either bare or as part of complete kits. New aluminum cylinders plated with a special Nikasil process, coating the inner surface using a special plating process, are supposedly available, but quite expensive.

Smaller-than-stock bores would be required to permit use of the Corvair in the under-2-liter class (less than 122 cubic inches), even with the early crankshaft. Salih Engineering may have provided these smaller cylinders on special request. Smaller cylinders would have to be notched out inside to clear the valves and it might be necessary to change the method of head gasketing. Pistons must, of course, would have been custom made.

If you are using oversize pistons and will utilize stock cast-iron barrels to enlarge your displacement, after the cylinders are bored scrub them thoroughly with hot detergent soapsuds and water—inside and out. Apply a thin coating of flat black paint or black stove paint to the exterior surfaces of the cylinders. This provides the best possible heat transfer from the cylinder fins to the cooling-air blast over the long haul. Actual tests of this black-paint idea show that it can reduce cylinder running temperature at the hottest point by as much as 80° F. Thus your oil has less cooling to do, giving your engine the added benefits of increased bearing life. A brand new bare iron cylinder might exceed the heat dissipation of the thin black coating, but rust could insulate the surface in short order. Incidentally, avoid any paint claiming insulating qualities, such as VHT Enamel or Hot Paint, both of which are perfect for other applications and especially for exhaust components. In this instance, we want to let the heat out of the cylinders, not keep it in.

10

Lower End

Corvair's sturdy lower end is built on the solid foundation of a two-piece permanent-molded crankcase. It is machined as one part, i.e., the halves cannot be interchanged with those from another crankcase. GM engineers were especially careful to eliminate distortion and maintain main-bearing roundness through attention to web design and by placing the crankcase bolts away from the bearing bulkheads. Two graphs from SAE Paper 140C show that main-bearing clearance "grows" as temperature increases. For this reason, the main bearings are fit closely on these engines, even when set up for racing. Thrust is taken by one half of the rear main bearing. One unusual design point—the camshaft has no separate bearings. It runs directly on bores in the crankcase, and is mounted below the crankshaft, just like a VW or Porsche. Unlike its German cousins, however, the Corvair cam has common cam lobes on the exhaust only. Another similarity is found in the webs from the center split to the valve lifter bores: the webs separate the crankshaft from the oil sump. You can find several such duplications when you compare closely.

Oil passages are internal within the case halves, with the main oil gallery on each side supplying the hydraulic lifters in the half. The oil pump is mounted separately in an accessory-drive housing that includes integral passages that connect the pump outlet to the oil filter and oil cooler and with the main oil galleries in the crankcase. The 1961 Shop Manual has the best coverage of the oil routing in the block.

Only minor changes were made in the crankcase after introduction, but the 1964 and later models included cast-in recesses to clear the added connecting rod radius caused by the additional stroke.

Crankcase Modifications

When modifying your Corvair it is a good idea to deburr the crankcase halves while you have them apart. The numerous sharp edges that are left by the factory do nothing to detract from performance, but they can definitely cause annoying cuts when you are working on the case halves. Take care not to scratch the bearing bores, the cam bearings, or the lifter bores. Special attention should be paid to the holes between the lifters and the main webs.

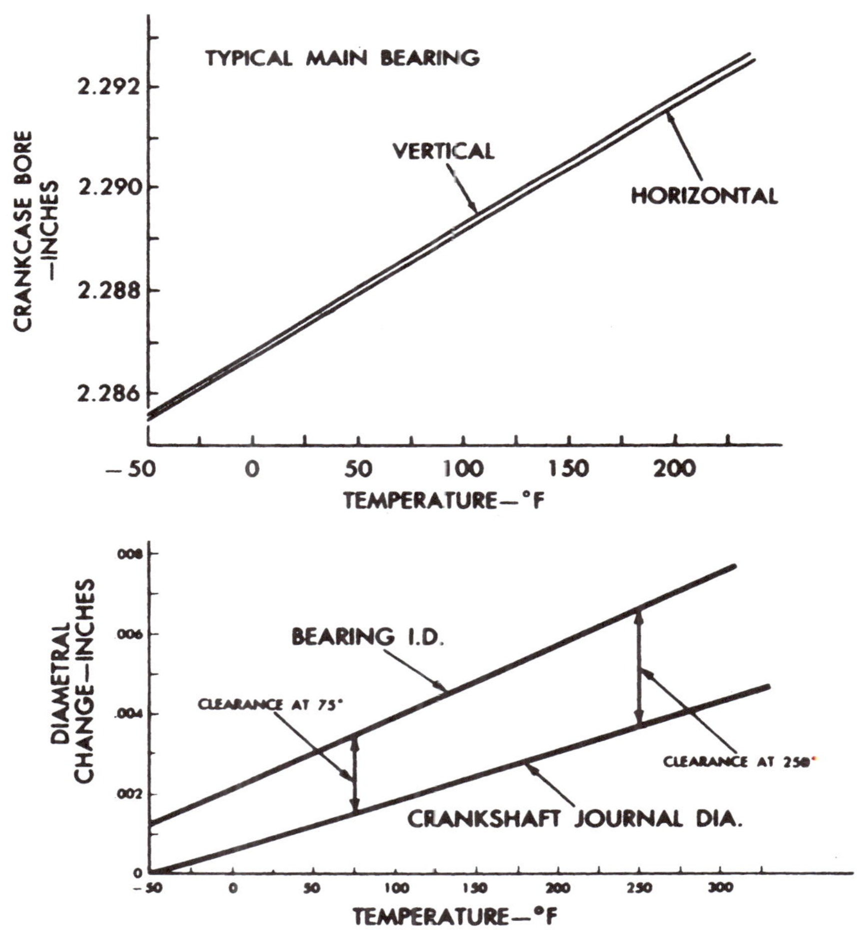

SAE Paper 140C provides graphs of crankcase-dimension changes with temperature. Top graph is crankcase bearing bore vs. temperature. Bottom one is main-bearing clearance vs. temperature.

With the late stroke, the rotating assembly is a tight fit into the cases.

Widened main-bearing oil-supply grooves improve oil flow to the bearings. Main-bearing shells are modified by adding one extra supply hole to each bearing half. This '61 case has been milled out ⅛ inch for rod clearance with the 2.94-inch-stroke crank.

These holes are often filled in with casting flash, which retards the flow of oil back to the case. By the same token, the three drain holes just under the case can be opened up somewhat to improve the drain-back situation.

A new alternative for more highly stressed engines is a new design, heavy duty, case stud kit from ARP fasteners. The kit is available from several Corvair vendors. It is not a simple bolt-in, but involves performing a regular case assembly and torquing of the original bolts, a slight reaming of the case halves' holes, one by one, to remove any play from the interface, then installing these tight-fit studs into the halves, then a complete disassembly, cleaning and reassembly. The payoff is a tight fit and better support for the crankshaft and camshaft.

One modification you can make will increase the oil supply to the main bearings. Grooves behind the main bearings are approximately ⅛-inch deep and 0.100-inch wide. They can be widened to 3/16 inch, offsetting the groove to one side to favor the entry of oil from the supply hole leading to the main oil gallery. This is not adding any groove in the actual bearing face. It is behind the bearing to help the oil flow into the bearing. The depth need not be increased. Use a ⅛- or 3/16-inch round-nosed rotary file in a small drill or die grinder to widen the grooves. I use masking tape as a guide for the increased width. Keep the offset in the same direction in each half so that the grooves will line up when the case is reassembled. Also make sure that no high spots are left to keep the bearing shells from seating into their bores. Clean the cases again after this operation. USE STOCK CLEARANCES.

Several articles have been written indicating that the oil holes from the galleries to the main-bearing bores can be increased to a maximum size of ¼-inch diameter. Before drilling into the case, check to make sure that this size drill will not break through or seriously weaken the oil passage wall. If a drill does break through, the defect can be repaired by heliarc welding.

If you are installing the 1964 crankshaft into a 1960–63 crankcase, recesses will be required for connecting-rod clearance as described later in this chapter.

Bearings and Bearing Modifications

Various bearing materials have been supplied in Corvairs, beginning with a steel-backed lead-base babbitt with sintered interlayer between backing and babbitt. This type was used in standard models through 1963. 1961 high-performance models had copper-lead alloy bearings on rods only. In 1962–63, the high-performance models had copper-lead alloy for both the rods and the mains. 1963 turbocharged models had aluminum rod and main bearings. In 1964, all Corvairs were equipped with aluminum bearings: Morraine M-400 on the mains, and M-500 on the rods. Only the aluminum bearings are supplied as replacements. When Corvair bearings are to be replaced, one bearing type is especially deserving of your interest: Clevite CL-77. These are widely used in racing and aircraft engines. At 7,500 psi rating, these CL-77s have a 25% higher load capacity than the Morraine 400s. Part Numbers for CL-77 main and rod sets are MS2584P and CB584P. When you remove rod bearings, you can check on the backside of the bearing to see if undersize bearings have been installed. Undersizes are indicated by the part number on the bearing.

An improvement in main-bearing oil supply is accomplished by adding additional oil-injection holes to the holes and notches already existing in the main bearing halves. Make these holes 3/32-inch diameter and carefully deburr the front and backsides of the bearing shells. One additional hole can be added to the flanged half of the rear main and two to the right half of the rear main. On the front and intermediate bearings, one hole can be added to each half, halfway between the existing hole and the notch. This will give five holes in the front three bearings, and six and a half in the rear main. USE STOCK CLEARANCES.

Crankshaft

At 28 pounds the Corvair crankshaft was the lightest U.S. automobile crank. Its low weight is made possible because the opposed-six design does not require counterweights—it is inherently balanced. Each pair of journals is spaced at 180 degrees to each other, and pairs are located at 120 degrees from each other.

Forged carbon steel was used for all 1960–63 standard and high-performance cranks, with Spyder models ('62–'63) forged in SAE 5140 chromium alloy steel. Spyder cranks are surface hardened by nitriding for increased journal wear resistance and improved fatigue strength in the fillets. GM lab tests showed Spyder cranks five times more resistant to cracking than the plain carbon-steel models.

All 1960–63 cranks had a 2.6-inch stroke. 1964 saw alloy-steel cranks used in all models and the stroke raised to 2.94 inches. Concurrently, a crankshaft torsional vibration dampener was included on all except the lowest HP models. Although some racers remove the dampener to reduce rotating weight, the stress on the crank can cause cracking and breakage. Most racers then run a dampener on the replacement crank. I highly recommend running the dampener. The dampeners are subject to deterioration with age. The rubber bond between the inner pulley and the outer ring can fail. In its first stage, this failure is indicated by some small vibration. Following that, the timing of the engine, while not changing, becomes impossible to set reliably, because the outer ring is turning in relation to the center. Finally, when the outer ring separates and leaves it usually takes the oil filter and the fan belt with it. This is a real mess. The easy condition check is to verify the matching stamped marks on the inner and outer rings. The stamping mark continues across the rubber filler, stamping the tiny "V" into both steel pieces. If the bonding lets go, the marks separate radially. This means it is time for a replacement, although they are being rebuilt. New dampener assemblies are now available from the parts vendors, these have all new heavy duty parts. There are tools available to remove the dampener assembly with the engine still in the car and the motor mount in place. This applies to most years, depending on motor mount configurations.

At the other end of the crank, the steel timing gear and flywheel flange is a single part that is key located and pressed onto the forward end of the crank with great force. Because it is not just driving the cam gear, but also driving the car, the gear is under a great load. It is not above spinning on the crank end and shearing the key. Because this results in the instant reorientation of the cam, it often results in bent valves as well. The failure mode for this is not necessarily visually obvious, although sometimes the gear will split. Often it will rotate a bit, then gall in place. Because it shears off the top of the woodruff key and takes it with it, the key to key relationship with the cam key may still look all right, viewed from the end of the crank. The actual crankshaft will be somewhere else, though. For racing applications the gear can be welded or match drilled and pinned in place. On certain engines, notably the 140 HP motors with air conditioning, this gear was modified to retard the camshaft slightly. The keyway in the gear was relocated to provide this feature. If you are building a motor with an unknown sourced crankshaft, make sure it doesn't have this "special" gear. The drive gear uses a non-standard thread to attach either the flywheel or the automatic flexplate. The threads are 11⁄32 inch – 24. Aftermarket gears are available with larger ⅜-inch flywheel holes already tapped. As an option, the original holes can be drilled and tapped oversize. An easy-to-use tool is available to facilitate the redrilling. Inexplicably, both the stock flywheels and the flexplates were already drilled for the larger ⅜-inch size. A bronze gear and a conical steel eccentric at the rear of the crank drive the distributor/oil pump shaft, and fuel pump, respectively. Don't forget the small spacer that properly locates the gear and eccentric. As you bolt the vibration dampener in place the bolt and the hub of the dampener clamp down and hold the complete assembly in place.

Crank Modifications

Cranks can be stroked, the main journals grooved for added oil supply to the rods, and the oil holes in the crank rounded off at their outlets (as even the new stock cranks have sharp edges that can wreck a set of bearings during assembly). The need for stroking Corvair cranks was practically eliminated with the introduction of the long-stroke '64–'67 crankshaft. These parts give more stroke than was practical as a modification to the 2.6-inch cranks ('60-'63) at a much lower cost. Because the Spyder cranks are nitrided, try to find Part No. 3833022 for a '64 Spyder, or '65–'66 Corsa. The crankshafts that have gone through nitriding are identified with a stamped "&" on the end. This treatment is removed with regrinding, so an undersized crank may no longer have the effect. The 2.94-inch stock '64 and later stroke is almost maximum in keeping with adequate wrist-pin-boss strength in the piston, piston guiding, cam-lobe width, and crankcase-wall thickness. Do not have the crank ground undersize to add clearance, unless damage has occurred, but polishing is okay. Use stock clearances throughout.

Installing a Stroker Crank (Including the '64 Crank) in Earlier Cases

The Corvair crankcase is tightly packed. When stroke increases are made, clearance must be added or the crankshaft cannot turn. This additional "space" within the case confines requires work on the cases, camshaft, rod caps, and pistons. Rod-cap recesses are made in the cases as shown in the drawing and photos, plus notches in the bottom of each cylinder. Notch and recess are placed together during assembly, and ring gaps are spaced to miss the cylinder notch when installing the pistons. The easiest way to provide the required clearance between rod caps and cam lobes is by using the 1964 cam and rods. If an earlier cam and rods are used, the rod-cap ribs must be ground at an angle, and the edges of the interfering cam lobes ground back to reduce their width. The later rods are a guarantee of better material.

When stroking a motor, piston compression height (c.h.) must be reduced, that is, the pin must be moved toward the piston crown. The reduction in compression height will equal .170, half the stroke increase. Example: The 1964 crank has a 0.340 stroke increase over the 1960–63 cranks. Half of that is 0.170, Stock '60–'63 c.h. is 1.76 – 0.170 = 1.590 inches, the new compression height. If 1.59 c.h. pistons are not available, a 0.170-inch thick stroke-spacer ring can be placed under the cylinder to move the cylinder and head out to compensate for the longer piston travel. Use of spacer rings requires longer pushrods to retain correct rocker-arm geometry as discussed in the "Engine Assembly" chapter. When using the 1.76 c.h. pistons, a 45° flat milled on the pin-boss undersides provides connecting-rod-cap clearance.

Before replacing the flywheel housing, remove the crank seal. Place a 0.065 to 0.070-inch thick shim under the seal when you reinstall it to move the seal rearward to mate with the sealing surfaces on the revised crank hub. This is only necessary when installing a '64 crank with a '60–'63 flywheel housing. The shim is not required when using a 1964-up flywheel housing. A 1964 stepped flywheel, clutch,

Detail of 3/8-inch relief for rod clearance: 60-63 cases with 64-65 cranks

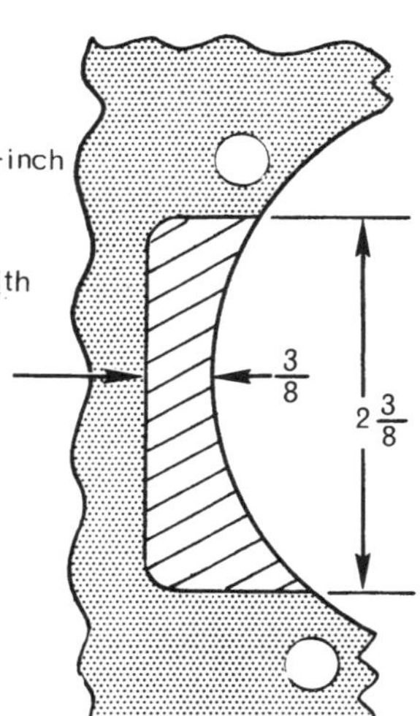

'64-'65 crankcases have cast-in reliefs for connecting-rod clearance with the 2.94-inch stroke. Camshaft is also modified for clearance in these later models.

improved "short" throwout bearing, bearing fork, and ball stud can be installed into the '60–'63 flywheel housing. There is adequate clearance. See also flywheel-bolt information in the "Clutch and Flywheel" chapter.

Even with the original dowel pins installed in the crankcase halves, the case halves can walk under competitive usage causing crankshaft and bearing problems. As mentioned before, in response to this problem, a new part was designed for the Corvair. These oversize case studs require match drilling of the case halves during assembly, but the vendors have some ingenious ways to deal with the concern, including offering tool kit rentals.

Aftermarket Crankshafts

Currently, there are no aftermarket crankshafts on the market. Some were offered but a couple of dimensional concerns stalled production. But others are in development. Within the limits of crank to cam interference, the new crank may be able to offer increased stroke, a happy addition to the limited cubic inches of the Corvair motor. Remember that the newest OEM crankshafts for the Corvair motor are about 50 years old and most have many, many miles—and shock loads—on them. Carefully examine any crankshaft that you will be investing the time on preparation and installation.

Connecting Rods

Steel forged rods are similar to other connecting rods except that there are no "spit holes" for lubricating the cylinder walls. Horizontal cylinders get adequate oil, even when starting. Three rod designs have been used to date: Type I, 1960–63 standard and high-performance models; Type II, 1962–63 Spyders with increased column cross section to handle added loads imposed by turbocharging; and Type III, 1964 and later (all models) using the Type II cross section with a modified cap to provide clearance for the rotating components. Type II or Type III (1964) rods should be used in any modified 1960–63 engine, with '64-up rods being the correct choice if new ones can be found. There is a kit out that replaces the Corvair connecting rods with modified VW Rabbit Rods. The difference in end-to-end length means that a different specific piston must be used as well. This is one of those racing-only mods now, but as the stock parts become harder to find, will become a real alternative.

Connecting Rod Modifications

An improvement in strength can be had for all-out racing use by shot-peening the rods. This service is usually offered in large metropolitan areas by aircraft-overhaul facilities. Rod side clearance, measured between the rod and crank-throw "cheek" should be 0.010 inch to 0.015 inch. The latter figure is preferable for an all-out racing engine, the former for a road job. More than 0.015-inch side clearance is not desirable because the oil throw off becomes excessive, making it difficult for the rings to provide oil control. Measure at one side of the rod, with the opposite side held firmly against the crank throw. Gain the additional clearance needed by clamping the rod onto a spud in a lathe and turning off the required amount.

The Corvair rod bolt was a 5⁄16-inch bolt, unique to the Corvair, although some 289 Ford motors also used a 5⁄16-inch bolt. ARP makes a replacement bolt set for the Corvair, it includes special nuts to clear the rod cap recess.

A billet crank is cut from a solid chunk of steel. Here is the rough cut on a special mill.

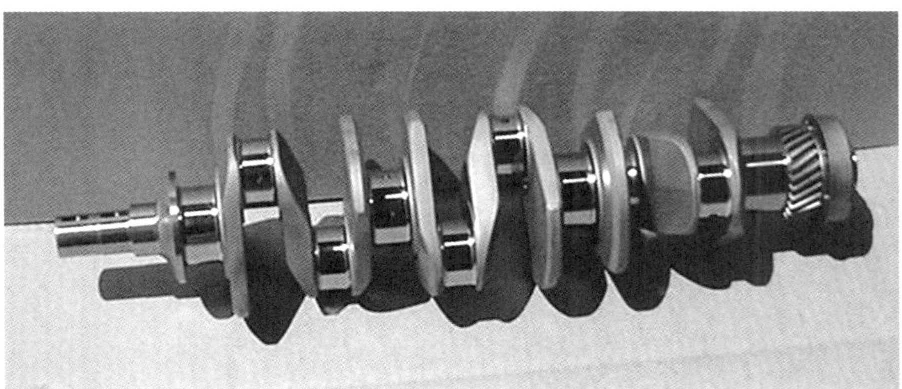

The final unit is strong and beautifully finished. The billet unit is probably stronger than an original forged crank. New alloys and material treatments account for this.

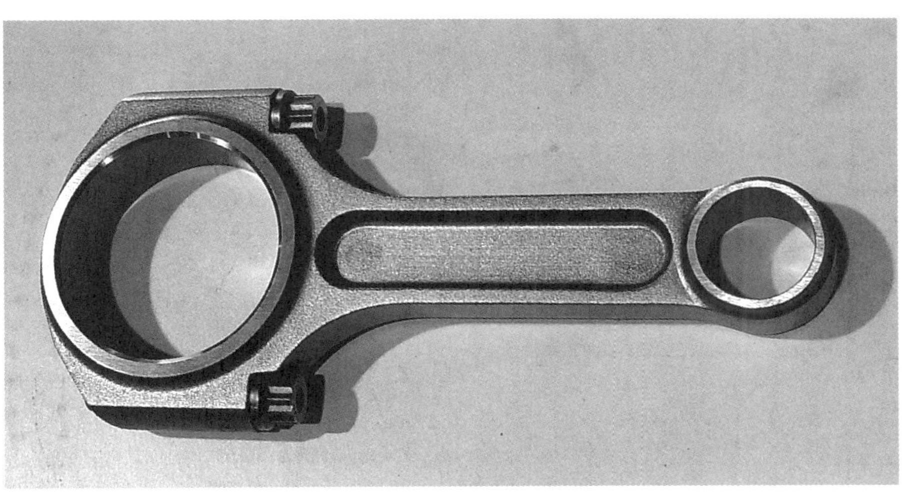

There are now several replacement forged rods on the market for the Corvair. Note the 12-point clamping cap screws are torqued from the piston end of the rod.

Larger-diameter rod bolts can be used when building an all-out racing engine. 11⁄32-inch-diameter Chevy V8 rod bolts and nuts (replacing the stock 5⁄16-inch) can be installed by using special fixtures and reamers. It is not an easy task, even though it looks simple. The most common failure point for a rod is the machined pad where the head of the rod bolt clamps the shell together. In order to install the larger 11⁄32-inch bolts additional material must be removed. This can weaken the rod as a whole and might result in a rod failure, rather than a bolt failure. The engine is just as dead either way. There are aftermarket connecting rods available from several sources in the correct size as replacements for the standard rod. They are, however, quite expensive.

Balancing

A number of different parts of the engine have now been discussed, with specific details as to their modification for improved performance at high RPM. By now you may even be eager to bolt the parts together and check out your efforts on the road or in competition. But, before proceeding with the assembly of the engine, it is important to get certain parts dynamically balanced. In fact, it is so important that you should budget the necessary money that is required, even if it means waiting for a few more weeks to add some external component.

Corvair engine parts are balanced by the factory, within certain tolerance limits, so that they will provide "acceptable" performance when the engine is bolted into the automobile on its flexible engine-mounting pads. But, even if you choose to build your engine with all new stock parts, there is that tolerance of manufacturing which can definitely be improved upon when the engine is to be operated at high RPM quite frequently or even on a continuous basis. This will be obvious from a close study of the accompanying photograph that shows an assembly from Bill Fisher's engine before assembly for installation into a Porsche (which explains the unusual looking flywheel and clutch assembly). You will note that each part is brand new, but grinding of the rods was necessary to make them match up with total weights equal, weights of all the big-ends equal, plus all the weights of the small ends. Two small areas were drilled out of the crankshaft webs. Some of the pistons were lightened to equal the lightest piston by turning off the balancing bosses under the wrist pins. The clutch pressure plate required welding on a weight to make it right, while holes were added to make the mass center of the clutch/flywheel assembly the same as its physical centerline. Assembly of this engine with the all-new parts would have drastically reduced its eventual life had balancing been skipped as a non-essential effort.

A smooth-running engine makes the automobile a pleasure to drive, but it is not your personal pleasure that is most important in this instance—it is the life of the engine. Balancing these parts adds significantly to engine life. Here is why. An out-of-balance crankshaft not only shakes the engine, it also hammers out the bearings because it constantly tries to rotate about an axis that is not the same as its centerline drawn through the main-bearing journals. At 6,000 RPM the 28-pound crankshaft will produce a 64-pound force if its mass center is offset a mere 0.002 inch. Or, a one-ounce imbalance at the outer edge of the flywheel will cause a 350-lb. force. If connecting rods or pistons are not balanced they will also apply uneven forces to the crankshaft, making it tend to run in an axis other than that through its physical center. The result is reduced life, parts fall off or break due to vibration-imposed stresses, and power is wasted through excessive friction and heat at the bearings which are working to keep the out-of-balance crank in an axis where it does not want to run.

The opposed-six design of the Corvair is often said to be "perfect" or inherently balanced. This does not mean that the parts do not need to have the dynamic-balance work done to get the crank mass centered perfectly, it merely means that the forces imposed by the pistons and rods cancel out at all positions so that no "rocking couples" are developed which would try to rotate the entire mass of the engine around its center.

If you live in a remote area where there are no balancing facilities nearby, box the parts and send them to the nearest major city where there are such facilities. Do not overlook doing this prior to starting the assembly of your engine.

Lower-end assembly from Bill Fisher's Corvair-engined Porsche. Arrows indicate where balancer has had to remove metal to get these new stock parts into static and dynamic balance. Note differences in rods; washer welded to pressure plate assembly.

11

Engine Assembly

Putting the Corvair engine together in a winning combination is absolutely tedious! If you are in a hurry, then bury this book or burn it because it will have been of no value you to you. It amazes me how many owners will put literally hundreds of dollars into reworking cylinder heads, improving the camshaft, increasing the displacement and so on and then insist on hurrying through the assembly of these expensive parts. Perhaps the most unfortunate part of the story is that these are the first "experts" to blame the part manufacturers for turning out such "lousy work." The old adage that haste makes waste can be no better applied than to the job of assembling a high-performance engine.

This chapter is probably misnamed as it will cover a great deal of information on preparing for the job at hand to reduce the frustration of putting together what is to be a high-performance engine at the conclusion. Some additional details are provided on torque measurements and clearance recommendations for handy reference. Careful attention to details ahead of the time when you pull the engine can definitely shorten the amount of time required to get it back together and into the automobile. Such items are important whether the car is used for competition where it is important to get to the next race session, or for transportation where the car should not be kept out of service for a lengthy period. Many auto enthusiasts have difficulty getting all of the needed parts together at the right time so that the work can proceed rapidly without delays. Order all of the components requiring time for delivery, then wait for these to arrive before taking the engine out of the chassis. Oil coolers, pistons, rings, bearings, camshaft, crankshaft, valve springs, retainers, big bore gaskets, and other components may take some time to arrive after you order them. Even when stock Chevrolet parts were available from the dealer, they were not always in the local parts bins and a wait for many of the parts was expected, especially where Spyder parts were involved. Now that Chevy dealers handle virtually no parts, you will be ordering from Corvair aftermarket vendors, as well as the few part manufacturers that actually know Corvairs. If you are using the 1964–65 crankshaft with the '64–'65 rods in an earlier 1960–63 engine, the parts can be dynamically balanced and the rods assembled to the pistons and aligned prior to disassembling the older crankcase. Today, the later ('64-up) core motors are more readily available than they were during the manufacture of the car. Unless you want to retain that "stock" ID number for your show car, the larger, 164 CID motor, is the place to start. Cylinder-head work is another time eater and you may want to rework a set of heads from a junked engine to reduce delays from this source. Some fellows even buy a complete used engine and either install it while reworking their "good" engine, or rebuild it instead and swap it out. You will be replacing or rebuilding most parts anyway.

Much of the special detail for assembling the Corvair into a high-performance package has been carefully covered in the preceding chapters of the book, especially those on cylinder heads, pistons, camshafts, and lubrication. When these previous recommendations are combined with an intelligent application of the details in the Shop Manual, you should end up with an engine that provides outstanding performance for its size. Let's look at the items that you should assemble before working on the engine, in addition to the usual hand tools:

Tubes are used to hold the cylinders in place during assembly. Without them, the crankshaft could not be turned to position the rod throws for connecting-rod installation.

1. Corvair Shop Manual(s)—two are required for most years:
   (1962/3/4 require 1961 manual
   1966/7/8/9 require 1965 manual)
2. Torque wrench, ⅜-inch drive, 0–600 inch-pounds
3. Dial indicator, 1-inch travel with suitable holding devices
4. Degree wheel
5. Ring compressor, preferably take-apart type as illustrated previously

6. Steel tubing, ½-inch OD, 6 lengths 4-¼ inch, 6 lengths 3-½ inch for holding down the cylinders during assembly
7. Two 2-inch lengths of plastic/rubber tubing to protect crankshaft journals as connecting rods are installed—commercial versions are available.
8. Piece of child's modeling clay for clearance checking
9. Strip of Plastigage, Type PO-1, 0.001 to 0.003 capacity
10. Permatex Anti-Seize Compound No. 404-D, one-pound can
11. Lubriplate, High-temperature, #105 Assembly Lube, 10-ounce tube, for rear housing and flywheel housing seals, plus initial lube of heavily loaded parts such as the rocker-arm balls or HRL Assembly Lube, Molybdenum
12. Permatex Aviation Form-A-Gasket No. 3 for sealing gaskets and crankcase halves
13. Carburetor and parts cleaning fluid, 4-gallon pail with parts basket

You can see that it might be difficult to round up these pieces on a Saturday afternoon. For instance, the Permatex 404-D is available through your local auto parts house, but they will probably have to order it for you. The smaller ⅜-inch drive, 0–600 inch-lb. torque wrenches are ideal for Corvair, especially for the connecting-rod nuts. Most ⅜-inch torque wrenches do not have a ratcheting tip. Dial indicators with 1-inch travel are available in many hand tool catalogs. For occasional usage I recommend Harbor Freight tools. Most are not of a professional grade, but if you are building a motor every five years, they will do just fine. Several firms, Moroso, etc. make degree wheels that will work on the Corvair motor. Corvair Shop Manuals may still be available from Helm, Inc., but the Corvair vendors have a quicker response.

Cleaning the engine parts is the first place to begin when you are disassembling the engine. After several years experience with Corvairs and other aluminum engines, I can assure you that money must be invested in a pail of carburetor parts cleaner. There are several ecologically friendly solvents now available. Most are actually water based. Steam cleaning will not get the job done, and the "hot tank" at most parts houses is for cast iron and steel—not aluminum. When the engine is all assembled, you will still own the cleaner and will be able to use it for your carburetors and other parts, or even mama's aluminum cooking utensils! The major portion of the dirt and grease encrustment should be taken off with solvent and a brush to avoid unnecessary dilution and pollution of your expensive cleaner, but the fried and baked-on dirt and varnish will only respond to the cleaner's potent action. You can fit the crankcase in the pail, a half at one time, soaking first one end and then the other. The same applies to the cylinder heads and this stuff does soften the carbon so you can scrape it off much more easily than you might imagine. When disassembling the motor, do not intentionally remove any pre-installed studs, such as head studs on the case halves or carburetor mounting studs on the cylinder heads, unless the condition of the stud demands replacement.

Installing the Cam Gear and Thrust Washer

The Corvair owner is fortunate because he is spared the worry of drive chains, which can stretch and plastic gears that can fail. The all-gear drive produces positive timing. Installed correctly, it stays right in time. Because the cam gear is aluminum and the cam is steel, they expand and contract at different rates. The only reliable method of assembly is to heat the gear and freeze the cam. However, installing the gear onto the cam is difficult, and must not be regarded lightly if you want the gear to have the proper clamping force on the end of the cam and stay on the cam "nose." The process, lightly skipped over in the Shop Manual, has been presented here in great detail because it is so very easy to broach the gear hole to a larger size during installation—then the gear will never stay tight without plating the nose of the cam to make it larger. Such tricks are

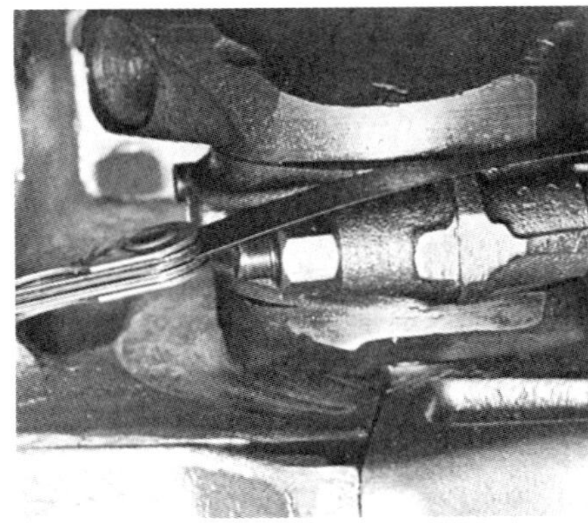

Rod side clearance is checked with a feeler gauge between the edge of the cap and the cheek of the crankshaft rod throw.

Permatex Anti-Seize compound should be used on every steel stud which enters aluminum. This is an essential compound for every Corvair assembler.

Close quarters! Assembling a Corvair, especially with the '64 or later crankshaft, requires careful attention to every detail. Note that the underside of the Badger pistons have been milled away at a 45-degree angle to clear the 1965 connecting-rod caps.

not only expensive, they consume more time and energy than would have been expended doing the job right in the first place. The first recommendation came from Bill Thomas. It does involve using a lathe for an initial press.

First install the thrust washer over the cam nose and then insert the cam key fully into its keyway. Coat the cam nose with a white-lead or moly lube, then place the entire camshaft in a deep-freeze chest. Heat the cam gear with a torch, on a gas burner, or in boiling water. You will need to wear gloves to handle the hot gear. Do not drop it or the teeth will be damaged. Place the hot gear in a three-jaw lathe chuck, working quickly so the gear does not cool off. Be careful not to distort the gear or its teeth with the jaws of the chuck, it needs only to be held lightly. Remove the cam from the freezer. Using the lathe tailstock, push the cam halfway into the gear bore, or farther if the tailstock is able to do so. Finish the installation in a hydraulic press, making sure that you continue to press the gear straight on. The proximity of an intake lobe to the cam bearing makes it mandatory to use a split holder on the cam when pressing the gear into position. There are no clearances involved, just press the gear right up against the thrust washer. Check the cam gear for run-out (wobble). It should not exceed 0.010 inch.

An alternative two-person method is to heat the gear on a stove in a heavy, flat bottom fry pan. No oil! The cam, as prepped above, washer/key/lube still goes in the freezer. Have your hydraulic press set to the correct height with a solid surface at the bottom. When you are ready, just place the frying pan, complete with gear, onto the surface of the press, hold the camshaft with two gloved hands to line it up, and have your assistant press it all the way through up against the thrust washer. The beauty of having your own big-enough lathe is the self-alignment that the tailstock should provide. If you are careful, it is a simple job.

The following steps, followed carefully, ensure that the camshaft will remain in place regardless of the stresses that may be imposed in competition. Degree the camshaft in the engine before you mill any holes, of course. Use a ¼-inch end mill. (Note, a drill press is not up to this job.) Mill two ¼-inch holes ⅞-inch deep half in the cam nose, and half in the aluminum gear. Enlarge the holes with a "F" drill. Tap the holes 5/16 inch-18 with a bottoming tap. Insert two Loctite-coated set screws into the holes. Screws should be ½ to ¾ inch long. Cam is now ready for installation.

Installing the Camshaft and Finding TDC

You will need a degree wheel such as that made by Moroso, a pointer, and a 1-inch travel dial indicator with suitable clamps and brackets.

Make sure that the crankcase halves are completely cleaned and that all of the modifications have been made as regards lubrication improvements. Be sure that the oil-gallery socket-head pipe plugs are installed in each crankcase half. Install the main bearings in each crankcase half per Shop Manual instructions, observing that the flanged half of the thrust bearing is to be installed in the left half of the crankcase. Install the crank on its lubricated bearings in left half of crankcase. Install camshaft into lubricated housing areas. Guide thrust washer into groove while indexing cam gear into crank gear so valve-timing marks line up as shown in the drawing. Install other half of crankcase but do not seal surfaces at this time. Install two of the eight through bolts and tighten them snugly. There is no reason to install all of the bolts because the assembly will be taken apart at the conclusion of the degree checking. I always mark the sides of the mating gear groove (valley) with the matching tooth, using machinist's blue dye or nail polish and scribe marks. This helps to ensure that the reassembly will be accomplished without errors.

Use the dial indicator to check that camshaft end play is between 0.002 and 0.007, with timing-gear backlash from 0.002 to 0.004 at one of the cam-gear teeth.

Install No. 1 cylinder, using spacer tubes, washers and nuts to hold the cylinder onto the case. Install piston and connecting-rod assembly. Top and

End view of camshaft that has been set-screwed as described in text. Ensures camshaft-gear retention under all conditions.

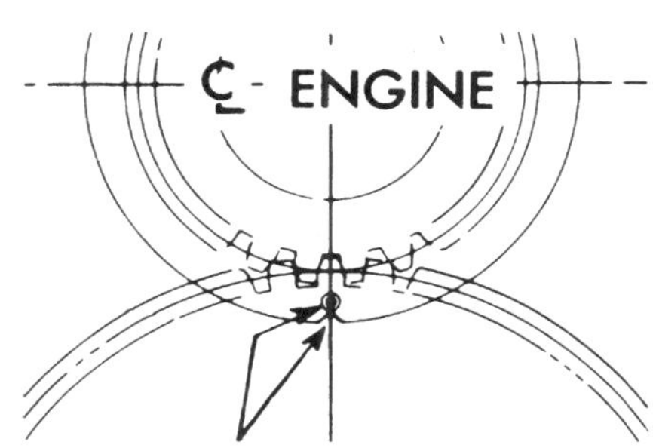

TIMING MARKS WITH #1 PISTON AT T.D.C.

bottom rings should be used to hold the piston squarely in the bore, particularly if a dial indicator is used to find TDC. Attach the dial indicator so that its stem rests on top of the piston. Rotate the crankshaft slowly in its normal direction of rotation. When the dial indicator shows that the piston has reached its maximum travel, attach the degree wheel to the crankshaft. The degree wheel can be attached with a bolt onto the end of the crank, or applied to the crankshaft pulley. Make an indicator tab from sheet metal or a piece of welding rod. Adjust the degree wheel and indicator to coincide with the TDC marking. Rotate the crank through an almost-complete revolution, stopping short of TDC when the dial indicator shows that the piston must still travel about 0.025 inch to reach TDC. Note the degree-wheel reading or make a mark on the crankshaft pulley. Continue rotating the crankshaft while the indicator passes through its maximum reading and backs down to 0.025 inch on the other side of TDC. Note reading. The ATDC reading should be the same as the BTDC one to show that you accurately located TDC. If not, start over again!

Another method allows positive location of TDC without using a dial indicator. It is so simple that it always amazes me to find how few engine builders are aware of how to do it. All you have to do is to bolt a strap across No. 1 cylinder. The strap should have a capscrew or nut in its center, placed to extend into the cylinder. Set the degree wheel and indicator finger at TDC by guesstimate. Rotate the crank by hand until the piston is stopped against the screw in the strap. Note the degree-wheel reading. Reverse the rotation and turn the crank until the piston again stops against the strap on the opposite side of BDC. Note the degree wheel reading. TDC will be exactly between your two readings.

This same method can be used when the engine is assembled and in the car if care is used. In this instance, the stop can made by modifying a sparkplug or just buying a long 14mm bolt. For the plug "tool," braze a piece of steel rod into an old plug base from which the insulator has been removed. The steel should extend about 1 inch from the threads. Remove all of the sparkplugs from the engine. Install your special "plug" into No. 1 cylinder. Rotate the engine by turning the crank-pulley nut with a box wrench in one direction with the fan belt in the other. Use extreme care so you do not damage a piston!

NOTE: *The factory's timing notch on the crank pulley, when used in conjunction with the marks on the accessory rear cover, cannot be relied on as a positive or accurate TDC location. Do not use it! There are no really reliable shortcuts for finding TDC or checking camshaft timing, so do not look for any. However, you can compare your accurate TDC location with the stock markings and then correct the stock marks to provide a true indication. For a permanent marking, clean off the area around the new mark location, then brush on about ½" of white paint on the outer cylinder of the dampener. After you determine the exact TDC, mark the matching point on the edge with a light chisel hit, or the sharp corner edge of a file, creating a small V notch, similar to stock. If the white paint is dry, you can take a toothpick tip of black paint and drop it into the V, then wipe off any extra black paint.*

Degreeing the Camshaft

All cam grinders have established their own quality standards to ensure that you will receive an accurately ground camshaft. They cannot control the other parts in your engine. Individual engines are different and their differences can and do drastically affect valve timing. You can obtain maximum performance by making sure that the cam installed in your engine opens and closes the valves as intended by the cam grinder. And, with the Corvair there is a second valid reason which makes cam checking essential: consider the hours and hours and hours of wasted labor if subsequent removal and a complete teardown of the installed engine is required to correct valve timing! Finally, the manner in which the stock timing gears are indexed leaves something to be desired. It is not any trouble at all to make an error in installation.

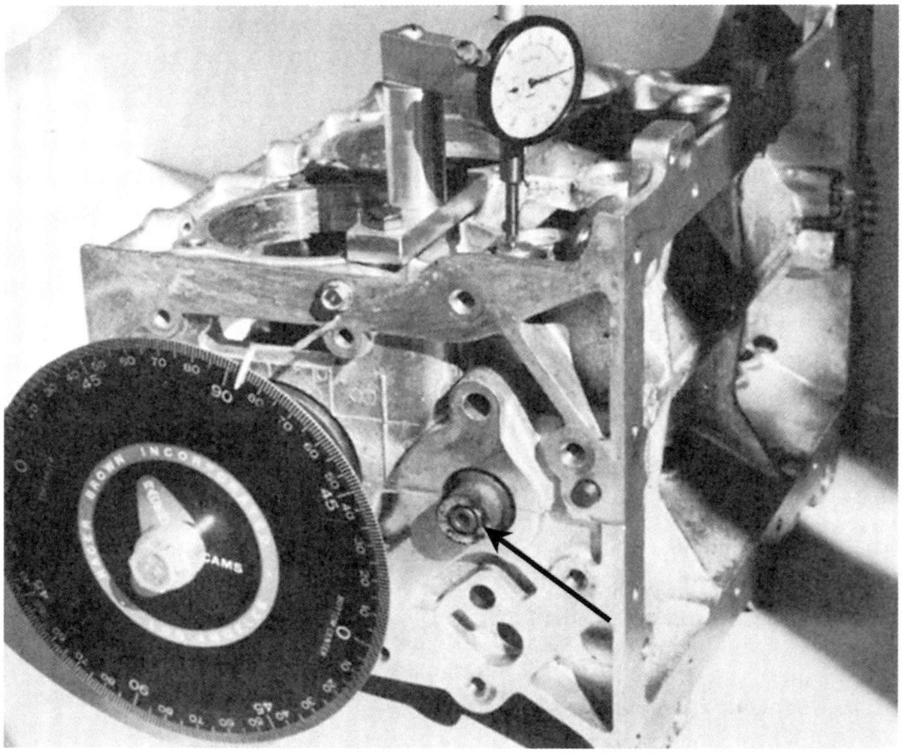

Valve-timing check setup. Dial-indicator finger rests on aluminum rod pressed into old tappet. Degree wheel is securely fastened to crank with indicator made from pointed piece of welding rod. Arrow indicates tach drive nut pressed into end of camshaft.

There is no doubt in my mind that you can install a stock or reground cam without checking. But, when you consider all of the tolerances working against you to cause less-than-perfect timing, why trust to luck? Some of the factors which affect timing include manufacturer's tolerances for the location of the keyways in the crank and camshafts and their respective gears plus the cam grinder's lobe-location tolerances. If, for instance, the location tolerance for the keyways in the shafts and gears is +0.75 degrees, you could get a tolerance "stack-up" of 4.5 crankshaft degrees total timing error with every item off by the maximum amount. If you decide to actually degree the cam, assemble the cam with a worn or opened up drive gear. You can even press a gear off an old cam, as long as the teeth appear to be in good shape. You want it to push onto your new cam tight enough to stay in place during the degreeing activity, but pull off with a little effort at room temp, when it is time to install the final key and washer and secure the gear.

It is assumed that you have carefully followed the instructions and have installed the degree wheel and accurately located TDC. Now a slight modification must be made to one of your old valve lifters (tappet). The lifter needs a flat surface on which the stem of the dial indicator can ride. A length of aluminum or steel rod can be fitted into the lifter body and a flat face turned on it. Do not attempt to use the radiused cup of the lifter as a contact point for the indicator stem as it can cause inaccurate readings and possibly damage the indicator. With the lifter on the heel of the No. 1 cylinder intake cam lobe, adjust indicator stem parallel to lifter travel in all respects. Preload the indicator about 0.010. Rotate the crankshaft and observe the indicator, watching for the point of maximum lift. Mark this in pencil on the degree wheel as "IN" for intake. Rotate the crankshaft exactly one turn in the same direction until the pointer again aligns with your "IN" notation. This places the lifter in the midpoint of the cam lobe's clearance section. The cam has turned exactly 180 degrees (half a turn). Without changing the indicator pre-load, set the dial indicator face to zero. Rotate the crankshaft in the direction of running rotation and observe the dial indicator. When it shows that the lifter has raised an amount equivalent to the checking clearance, record the degree wheel reading. This will be a certain number of degrees BTDC (before top dead center). If no checking clearance is specified or the cam tag has been lost, use 0.010 inch as this figure will be close for most camshafts.

Continue to rotate the crankshaft in the same direction until the lifter has risen up in its bore and fallen back again as it followed the opening and closing flanks of the cam lobe. Watch again for the indicator to reach the checking clearance or the arbitrary figure of 0.010 inch just mentioned. Record the degree-wheel reading as the number of degrees between the pointer and BDC. Add 180 degrees to your two readings to get duration of opening measured at the checking clearance.

Repeat the process several times to eliminate the possibility of errors. You may find that it is necessary to apply pressure against the lifter to hold it against the lobe, especially if you are making the check with the cylinders horizontal. I prefer to make the check with the No. 1 cylinder vertical so the lifter will fall back of its own weight. Each repeat check must be started at the "IN" mark on the degree wheel and the indicator's needle and dial face must be at zero. Cam lobes and lifter bores must be scrupulously clean and lightly oiled so the lifter can move in and out of the bore without binding and of its own weight. With everything right, the indicator will reach zero and stay there before and after the mark on the degree wheel is reached.

Transfer lifter and dial indicator to No. 1 exhaust and repeat the entire process, beginning with finding full lift, then turning 360 crank degrees and marking the wheel "EX," using this as your starting point, with the dial indicator preloaded about 0.010 inch and the indicator face set to zero. Exhaust opening checking clearance will be reached at a point BDC. As with the intake, the number of degrees from the BDC mark on the degree wheel must be counted to get the opening point. Record this figure. Again, rotate the crank and observe the indicator as it indicates opening and then closing back down to the checking clearance. This will occur at a point ATDC, and the number of degrees from TDC must be recorded. Exhaust duration is calculated by adding 180° to the two readings recorded for exhaust timing. Again, repeat the process several times to eliminate the possibility of error.

Overlap is easily determined by adding the intake-opening degrees Before Top Dead Center to the exhaust-closing degrees After Top Dead Center. The previously outlined procedure, under ideal conditions, will show that timing points coincide precisely with those shown on the cam tag. But, due to factors already stated, variations can occur. A difference of one or two crankshaft degrees between actual installed timing and the cam-tag specifications should be considered the maximum allowable error. A closer tolerance of +/− 1.0 degree at the crankshaft is even better.

If the timing check shows that actual timing is way off—14 degrees or more—then the timing gears are probably misindexed by one or more teeth. Correct such a situation and rerun the timing checks. Racer comments that if an engine is run in this condition, or if the valve-timing check is made at the valves instead of at the lifters, it is a foregone conclusion that you will end up with an engine full of bent valves at the least. And, very probably, you'll also get bent pushrods, damaged rocker arms, and possibly even a bent or broken camshaft. These are just a few additional reasons why the timing check should be made at the lifters, as opposed to checking the timing at the valves, or not checking at all. Checking valve overlap with feeler gauges at the valves by the so-called split-overlap method is not recommended under any circumstances because of the errors which can occur in rocker-arm geometry. Feelers cannot be used for accurate checks!

Correcting a timing error requires rotating the camshaft in relation to the crankshaft, which can be confusing. It is important to take care to advance or retard the camshaft and not the gear! Remember that the gears remain in the

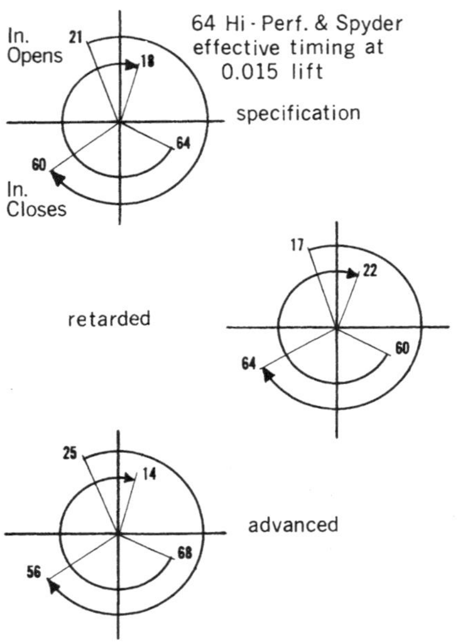

Using offset key to adjust valve timing

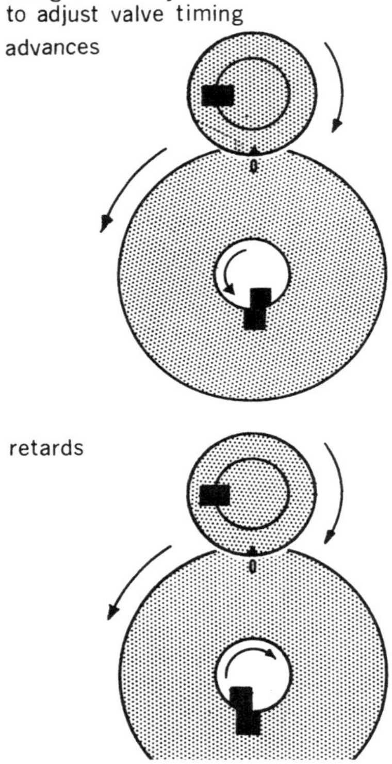

same relative positions, it is the shafts that must be moved. This is true in all instances except where the cam and crank gears are misindexed, in which case the gear/shaft assembly is rotated. Assuming that the timing gears are held stationary, the drawing illustrates the direction the cam must be moved to advance or retard its action. Remember that the cam is only moved one half degree to effect one crankshaft degree adjustment. Three valve-event drawings compare specification timing with advanced timing (all events early) and with retarded timing (all events late). Other sketches show how to use an offset key for corrections. When an offset key is used, repeat the entire degree check to ensure that the correction has been made properly, that is, corrected the right amount and in the right direction. A movement of 0.0049 inch at the cam nose adjusts crank timing one degree. Correcting a 5-degree timing error would require an offset key to offset the camshaft 5 degrees × 0.0049 = 0.0245. Sketches show installation. Inexpensive offset keys can be used for making corrections, but all are labeled in camshaft degrees. You will find that the keys move the Corvair cam slightly less than stated. Two that were measured gave corrections as follows: 2-degree key gave 3.7 crank degrees; a 4-degree key gave 7.5 degrees at crank. Rechecking and verifying cam timing after installing an offset key would be essential!

While you may be familiar with making these corrections at the crankshaft—and they are easier to make at the crank on nearly all engines other than the Corvair—the Corvair has its own peculiarities of design. The very heavy press fit of the crank gear onto the crank "nose" requires a heavy press and careful work to install and remove it, thus you should make corrections at the cam gear. A positive method for retaining the camshaft gear is discussed elsewhere in this chapter.

Rocker Arm Geometry

In addition to installing the camshaft in the correct relationship to the crankshaft, you must observe another important point of engine construction. This is the relation of the rocker-arm-tip radius to the valve-stem center. If this relationship is correct, with the centerline of the rocker tip coinciding with the valve-stem centerline at 50% of total lift—good things happen! Correct lift throughout the lift cycle of the cam is assured, side thrust of the valve stem against the guide bore is minimized to reduce wear and consequent oil consumption, friction is reduced, and higher RPM can be reached before valve float occurs. This is a point of construction that is often overlooked but it makes a vast difference in engine performance. While motorcycle race-engine constructors used and wrote about these things for decades, auto-engine modifiers have only recently come to the realization that extra HP can he gained through careful attention to all details. Care produces more HP than a trunk full of "black magic" tricks. Bike builders often had only one cylinder so they had to make it produce maximum HP. These fellows are sharp!

Getting back to the tip/stem relationship, there are several things in the Corvair engine that can affect this relationship: head milling, cam, valve-seat location, valve-stem length, cylinder length, lifter dimension, rocker-arm differences, and pushrod lengths. Some of these things become immediately obvious as sources of problems when we are working on the engine. Then we usually shrug our shoulders and forget about the problems and go right ahead with the engine assembly, never realizing that in so doing we have seriously shortchanged our engine's breathing ability. Fortunately, the ball-pivot rocker arrangement of the Corvair permits setting each rocker arm at the precise height to get a perfect tip-to-stem setting. Adjustments are possible through varying the height of the valve stem and/or the pushrod length; do not "sink" the valves. Another option is to install small valve caps onto the tips of the valve. This will build up the effective height of the valve tip and provide a slightly larger surface area for the rocker arm to move over. They are unlikely to come off in operation, even if you float the valves once in a while.

Before continuing, let's look at some of the other aspects of ball-pivot rocker arms. Although these have virtues of low weight, low manufacturing cost, and simple installation, they also present problems and must therefore be considered a mixed blessing. Critical examination with a stroboscope to slow the motion for observation shows that the ball-pivot rocker opens and closes the valve with a continuously changing pivot-to tip radius, moving through arcs of varying radii. With this changing situation, the rocker-arm ratio is highest when the pushrod is nearest to the pivot point, let's call it 1.57:1. It lessens as the pushrod moves away from the pivot, say 1.45:1. These figures, incidentally, are merely examples chosen to illustrate the problem and do not represent the actual ratios which may appear. The problem is still further complicated by manufacturing tolerances that produce variances in the distance from the ball seat to the rocker tip. Proof of the differences and strange behavior afforded by these arms can be shown without a stroboscope. Just measure the valve travel with the cam at half and full lift, using all 12 rockers, one at a time, on the same valve. Some will produce more than 50% expected travel at half-lift, while others will show the opposite to provide more than specified valve travel at full-lift with less than 50% at the half-lift point. Obviously, you want to get 100% valve travel with the cam at full-lift position.

Setup for checking rocker-arm geometry and pushrod lengths. Adjustable pushrod is two-piece construction to permit installation and removal. Pushrod guides are installed.

Check also requires making a special tappet with modified seat height as described in the text. Lightweight springs are used only for checking.

The following information has been written as the result of several long conversations with Chevy expert Frank McGurk. His method of checking and making the necessary corrections was developed for setting up winning stock-car engines for competition. It makes a lot of sense. Because of the production-tolerance stack-up which creates rocker-arm differences, the situation must be straightened out as follows: (1) tailor the individual valves so that their stem length is as nearly identical as possible, and (2) tailoring the individual pushrods to provide identical opening at full lift. This sounds complex and time-consuming. It actually is simple once the principles are understood, not really time-consuming, and you get free horsepower for your efforts!

Try to make all of the valve-stem heights (not lengths) identical within +0.005 inch (the rocker-arm-cover-gasket surface can be used as a reference point from which to measure heights). If any of the valves have to be shortened by grinding on the tips, reharden the ends by heating them cherry red and quenching in oil. Use a fairly large torch so the tip will get to cherry red quickly without the heat extending to the keeper groove. Hardening is only wanted on the tip. It is assumed that you have already seated the valves per directions in the "Cylinder Heads" chapter. Reseating will change the stem heights, so do any seating prior to establishing the heights.

Modify two tappets so that the pushrod seat will be raised 50% of the cam lobe lift in one (minus the amount of lifter pre-load to be used in the completed engine), and lowered the amount of the lifter pre-load in the other. For instance, if the cam lifts 0.300 at the lobe, make a button to raise the pushrod seat 0.150, minus the amount of lifter pre-load to be used. We will call them 50% lifters. Measuring pushrod-seat dimensions is easier when you use a 5⁄16-inch diameter steel ball in the seat. Measure from the top of the ball to the spherically ground lifter face which contacts the cam lobe. The terms "lifter" and "tappet" are used interchangeably in this book. With the cylinder heads torqued into place and a very lightweight spring under each valve to permit opening the valve easily with your hand, follow this procedure. Insert the lifter with lowered seat onto the cam-lobe heel and then add pushrod and rocker arm. Adjust rocker arm to zero clearance if using hydraulic lifters, to maker's specification if solid tappets required. Position a dial indicator over the valve-spring retainer, preloaded about 0.500 to record valve travel. Set indicator face to zero. Depress valve with your hand and remove the pushrod and "stock" lifter. Insert the 50% lifter and reinstall the pushrod. Record indicated valve travel. If the lift obtained is in excess of 50% of the cam grinder's total-lift specification, the pushrod is too long. If travel is less than 50% of total lift, pushrod is short. In either instance, a correction must be made if you are to get all of the performance the grinder put into your stock or special cam.

How much longer or shorter must the pushrod be made? One way to find out is accomplished simply by using an adjustable pushrod with the 50% lifter. Adjust the pushrod to provide 50% travel at the valve, then remove the adjustable pushrod and measure its length with 12-inch calipers. Record that length. Make a pushrod to that length and keep it with its own rocker arm. Identify both items as to their eventual location in the engine, i.e., #1 IN, # 1 EX, etc. They can be easily marked with layout blue, using a scriber to scratch the information through the blued area. Do not file notches for this purpose! Why not just use adjustable pushrods and save the bother of making special-length pushrods? First, the adjustables are not made for the Corvair. Secondly, they are too heavy and would require higher valve-spring pressures that would increase lifter wear, cam wear, and limit the engine's RPM capability.

There are no standard adjustable pushrods small enough for making this check, so you will have to make one from a stock pushrod as discussed below. Note the adjustment must be in the area between the head and the crankcase as the tub shape of the head would prevent simple adjustment just under the rocker arm as is the more normal situation with other engines.

Custom Pushrods–How to Check and Measure

Most Corvairs originally used one of two standard-length pushrods. They are a light design, perfect for the original application. As soon as modifications start, the pushrods will be the "wrong" length. The contact point of the rocker on the tip of the valve should start on one side, and sweep across the tip. Milling the heads moves the rocker to valve contact point closer to the engine centerline. With a standard-length pushrod, the rocker now has to lean over, changing the contact points of the rocker tip on the valve during operation. It might start in the middle now and sweep over the side of the tip. This is hard on the valve guide, because it pushes the valve sideways. It also causes ugly wear on both the valve and the rocker. If the valve seat is machined, the valve drops deeper into the head and sticks out higher for the rocker to hit. All of these functions add together and can result in the need for custom-length pushrods. It is easy to buy pushrods at the correct length—as soon as you can find the correct length! Clark's has a kit to calculate the correct length, you can also build an adjustable pushrod with a modified old pushrod, some threaded rod and some nuts. If you can, try to find a solid lifter (non-hydraulic) that is the same height at the hydraulic lifter in your motor. Replace the lifter in the cylinder you want to check and rotate the motor to a point that the lifter is not being moved by the cam lobe (zero lift). Cut the extra pushrod in half and take a couple of inches out of the middle. Put four nuts in the middle of a 5" piece of threaded rod and slide the cut tubes over the two ends of the threaded rod. Cinch two of the nuts tight. With a head bolted in place, less the pushrod tube, install a rocker arm, slide the adjustable pushrod into place. The piece that will contact the rocker arm has to slide through the guide from the head side. Now to change the effective length of the pushrod, just move the non-locked two nuts away from the locked pair. This will lengthen the pushrod. Changing the length relocates the starting point for the rocker. First check the point on the valve tip that you are contacting. You can relocate the rocker position up or down, with the center adjuster nut, to make the contact point on the valve the optimum point, near the side closest to the pivot ball. Once that is correct, adjust the pushrod length to take up the slack. At this point, you can turn the engine through a rotation and watch the path that the pushrod tip takes across the tip of the valve. The optimum pattern is the closest to the middle of the valve tip back and forth. You can move that path by varying the rocker ball nut and the pushrod length. Once you have the pattern you want, cinch up the other pair of nuts. Undo the rocker nut and extract the two parts of the "adjusted" pushrod and measure it. Clark's sells custom length pushrods in their "Otto Parts" line. Ray Sedman at American Pi and Smith Brothers in Oregon also supply custom lengths.

No stock pushrod will provide the needed accuracy unless you adjust the overall valve-stem lengths by stellite welding onto the ends (tips) and then grinding the tips to get the exact lift, or using the valve caps. If you ever worked on flathead Ford V8s, you will understand exactly how this is accomplished.

By now you are perhaps wondering just why so many engines run as well as they do. You can begin to understand why the top tuners always run out in front, but never ever brag about how quickly they can assemble a winner. Now that the need for careful assembly has been reemphasized, proceed to the next step, installation of the valve springs.

Always use the lowest valve-spring pressure that will cause the valve train to follow the cam profile to the desired RPM with some safety factor for over-revving. Excessive spring pressure merely hastens wear of the entire valve train. The cam grinder will suggest the springs and their installed height.

Case Half Assembly

Work on a clean, well-lighted table as if you were a doctor performing in surgery. Consider dirt the mortal enemy of your engine. Give close attention to using the torque wrench. Crankcase bolts are not tightened—just the nuts, and these should have flat washers under them, preferably at both ends of the bolt. The assembly of the crankcase, bearing, camshaft, and crankshaft unit is never a simple "just bolt it together" process. The two case halves change relationship to each other as you torque the original eight case bolts. Ray Longanecker of Say Ray Foreign Auto in Palo Alto was helpful in showing how one of these assemblies can go together. If you clamp it together with no special care, the crankshaft will probably be tight at several points in its rotation (with no rods installed). You must proceed slowly, taking up only a few ft. lbs. of torque at a time, and checking the crankshaft for free rotation as you proceed. Start at 10 ft. lbs., then tighten in increments of 10 ft. lbs. until you reach 40 ft. lbs., and finish off with 5 ft. lbs. increments until you reach the 55 ft. lbs. specification. You must check the crank for free rotation almost continuously. If or when you get to a point where the crank does not turn freely throughout its rotation, it will be necessary to beat on the cases with a large rubber mallet until the crank again turns freely throughout its rotation. You may get the cases almost completely tightened and then find that no amount of beating will move the case halves enough to get free rotation again. In this case, you back off the nuts and start over again. I have found that some engines will not stay "free" when the stock tightening sequence is used—a modified sequence is sometimes required. Chevy used a little bit of sealer at both ends of the case during assembly. This will prevent seepage around the oil pan and the top cover. That is the only amount recommended. The full case halves should not be coated. If you decide to use the special ARP studs and install them via the suggested reamer, you should proceed with the above procedure first. Only then will you be positively supporting the crank in the best location.

When you install the connecting rods, keep rags wrapped around the pistons and the rod shanks to avoid nicking either the pistons or the crankcase. Be sure to use short lengths of plastic tubing, or the commercially available boots, on the rod bolts as you slide these over the crank journals. Check the fit of each bearing with Plasti-Gage. It is an essential step. Clearances of the piston skirt to cylinder bore are rechecked prior to installing the cylinders. The fit of each ring to its groove is checked as the ring is installed onto the piston—after checking the ring end gap. The chamfer at the bottom of the cylinder can be used to install the piston and rings into the cylinder without a compressor. A ring compressor is not needed, but they are convenient. A take-apart type is pictured on page 63. There are several "take-apart" compressors available if you need to reinstall a barrel over an already installed piston/rod assembly.

Remember to install the thin steel gasket under each cylinder before it is inserted in the case. Use a light sealing compound on both sides of the lower gasket to ensure against oil leakage. Once a cylinder is installed, add a pair of the steel tubes (one long and one short) to hold the cylinder in place, using cylinder-head nuts to hold these onto the studs. When all of the cylinders have been installed, observe the clearances between the connecting rod caps and underside of the pin boss of the opposing piston. Also check the clearance between the rod cap and the opposite crankcase wall. There should be at least 1/16-inch clearance at both points. If, during assembly, you reach a point where there is any binding, do not proceed until you have corrected the condition.

You are ready to install the cylinder heads, but a final check is necessary to measure piston-to-valve clearance and to check clearance between the piston and cylinder head (on heads which have a flat deck). Old head gaskets are used and one head is prepared with valves, springs, and retainers in one cylinder. A strip of clay 1/8 inch thick and 1/2 inch wide is placed across the piston crown in a fashion to coincide with a centerline drawn through the two valve stems. Another strip of clay 1/8 inch square is placed at 90 degrees to the first strip to coincide with the deck area of the cylinder head. Torque the cylinder head into position with the pushrod guides installed on all cylinders, but with no pushrod tubes or O-rings. Fill the two hydraulic lifters with oil and install these in their bores, add the pushrods and rocker arms for the cylinder being measured. Adjust the rockers for no clearance by watching lifters installed for the opposing cylinder. When the opposing cylinder intake valve is full open, set the intake clearance on the cylinder being checked; when opposing exhaust is full open, set the exhaust. The engine is now simulating a complete assembly with the valves being pushed to their outmost position in the cylinder. Turn the engine through two complete revolutions of the crankshaft. If you feel any resistance against turning, stop and determine the cause. Use a small penlight to peer into the sparkplug hole to see if the piston is hitting the head or either of the valves are hitting the piston. If no resistance is encountered, remove the cylinder head after the two-turn procedure.

Remove the clay that was compressed by the piston and the deck of the head. Measure the compressed thickness. If the thickness is less than the 0.018" minimum, add additional gaskets under the cylinder. Do not use additional head gaskets. If the clearance is too great, which would provide drastically reduced turbulence, it could mean that you did not machine the "step" out of the deck portion of the chamber. Or, if the heads were properly machined, the rod lengths or piston compression heights could be off. In the event the heads are correct, the reduction in deck height can be

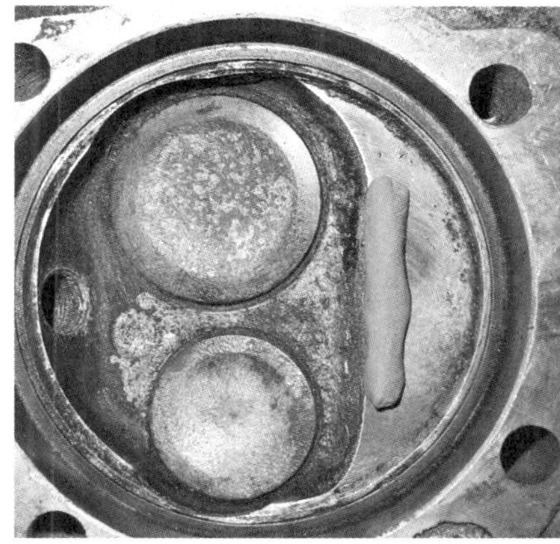

Roll soft clay and stick into chamber where piston will contact. Install head, and semi-torque it.

Crank engine through a full turn and piston will flatten clay.

Cut through clay in middle.

Measure thickness, either on the head or remove and measure it. System is essentially the same for measuring valve-to-piston clearance. See text.

achieved by turning off the base of each cylinder an equal amount. The cylinders will thus insert themselves farther into the crankcase spigots and deck height will be reduced. Note that the 0.018-inch deck height is a minimum for racing engines where the heads will be frequently removed and the carbon cleaned out. Road jobs have a greater clearance of 0.025 to 0.030, with 0.060 recommended for the pre-1964 Spyders. Any correction made to the deck height will affect the valve-to-piston clearance measured by the other strip of clay. Should there be an imprint of one or both of the valves on this other clay, carefully remove it and measure the thickness at the thinnest point. If the thinnest point is 0.090 or greater, there is no problem. But, if the measurement is less than that figure, the clearance must be increased on all pistons equally to maintain engine balance, even if the installed heights of the valves in the chambers are slightly different. In that respect, it will be necessary to measure the heights of the valves below the chamber surface (with chamber facing up) to determine whether any variances might affect the measurements made on the one cylinder. This will not usually be a problem on stock heads, but where large valves have been added, it must be done. If any valve of the same type (intake or exhaust) is closer to the gasket surface than the one that was checked, it should be checked with the clay method, or reseated to equal the others.

Clearance for the valves is added where necessary (usually on cams lifting 0.450-inch or more) by cutting valve reliefs or "eyebrows" into the piston. This is usually required only on the intake valves. Racer Brown's camshaft catalog contains a complete description of a simple way to machine these reliefs into the tops of your pistons. Even major relief, if done an all cylinders, should not affect engine balance. Once checks have been made and the corrections applied, engine assembly can be completed. It is best to do these checks when establishing pushrod lengths. Use copious amounts of oil during the assembly process. Remember that the first few revolutions of the engine contribute a lot of wear unless there is adequate lubrication. GM's E.O.S. (Engine Oil Supplement) must be used to coat camshaft, lifters, and in the initial oil fill. I always turn the oil pump with an electric drill until pressure registers on the gauge. This takes only a few minutes and ensures beyond any doubt that you have supplied lubrication to all of the parts. Similarly, the use of a moly-based lube on the rocker-arm balls is recommended to avoid galling-type wear of these items when the engine is fired up.

Installation of the Cylinder Heads

Once the "short-block" of the Corvair is assembled, it is time to prep and install the cylinder heads. The first step is to identify the left head and the right head. I have stumbled upon several motors that had the heads swapped left to right, fully finished and then "adapted" with extra hoses. It must have been embarrassing. Install the cylinder head gaskets into the chambers with the sharper, tapered seat outward, then slide the heads over the upper studs, then the lower ones. Align the tops of the barrels and slide the head on. Install the O-rings (only use Viton) in the outer groove, using the assembly lube to make it easier. Slide the other, shorter groove end of the pushrod tube through the hole in the cylinder head then install those O-rings. Slide the head over the upper studs and snug down the upper flange nuts—lightly. At this point, make sure that all of the lifters are installed against the cam. (Again, a little lube on the contact face of the lifter won't hurt.) Now you must install the lower air diffusers, using the springs to snap over the studs to retain them. The pushrod tubes can now slide into place. To properly seat the tubes, take the proper size ½" drive socket, slide it inside the outer flange of the tube and tap it with a mallet. It will pop down and squeeze both O-rings into contact. There are 12 more O-rings to install, though. Six on each head. The pushrod guide plates and the lower studs retain the final O-rings. GM stamped the guide plates with a large "U" on one side. That "U" should be visible to you after the plate is installed. This is not an option—the plate must be visible after installation. The slots used to align the pushrods are directional. If the plate is installed upside down, the pushrods will bend or break. Slide the pair of rocker studs through each plate, then install a lubed O-ring over each stud and screw the studs down snugly on the lower cylinder studs up against the head. When all six are done on both heads, you will have to torque down the cylinder head studs. This is not a fun job, but it must be done in a specific procedure to ensure proper sealing and engine longevity. The factory suggests a complex but effective pattern for the head, Richard Finch suggested that each cylinder be treated as an individual and the four studs torqued at a time. Either pattern requires switching back and forth between the 9⁄16" socket for the top studs and the 13⁄16" socket for the lower studs.

Lubrication, Cooling, and Breathing

Everything you do to increase Corvair power increases the production of heat simultaneously. To maintain engine temperatures at usable levels requires increasing cooling abilities while accomplishing other modifications. Because oil is used as an essential element in eliminating heat from the Corvair engine, lubrication has been discussed in this chapter as inseparable from air and oil cooling.

You cannot, except at great expense, increase the cooling-fin surface of the cylinders or heads. General Motors carefully engineered these items to provide maximum heat removal from the available cooling air consistent with the limitations of modern volume-production techniques.

You can greatly improve cooling by changing to a folded aluminum fin cooler 1960 to 1962 Corvair part, as long as you are a regular at keeping the area clean. The 12-plate cooler introduced on the last Spyder and carried on for all high-performance use, probably cools as well as the earlier, smaller folded-fin cooler, but is much less likely to get plugged up with debris. Some additional help can be supplied by improving the heat-radiating characteristics of the engine's cast-iron and aluminum surfaces, both finned and otherwise. A little improvement is possible through exchanging certain of the sheet metal parts for finned-aluminum components. Added cooling is also available through better use of the oil's heat-removal characteristics via increased flow made possible through bearing and oil-pump modifications. Adequate crankcase breathing is also important to keep engine temperatures at reasonable levels. All of the foregoing modifications are fully described in this chapter.

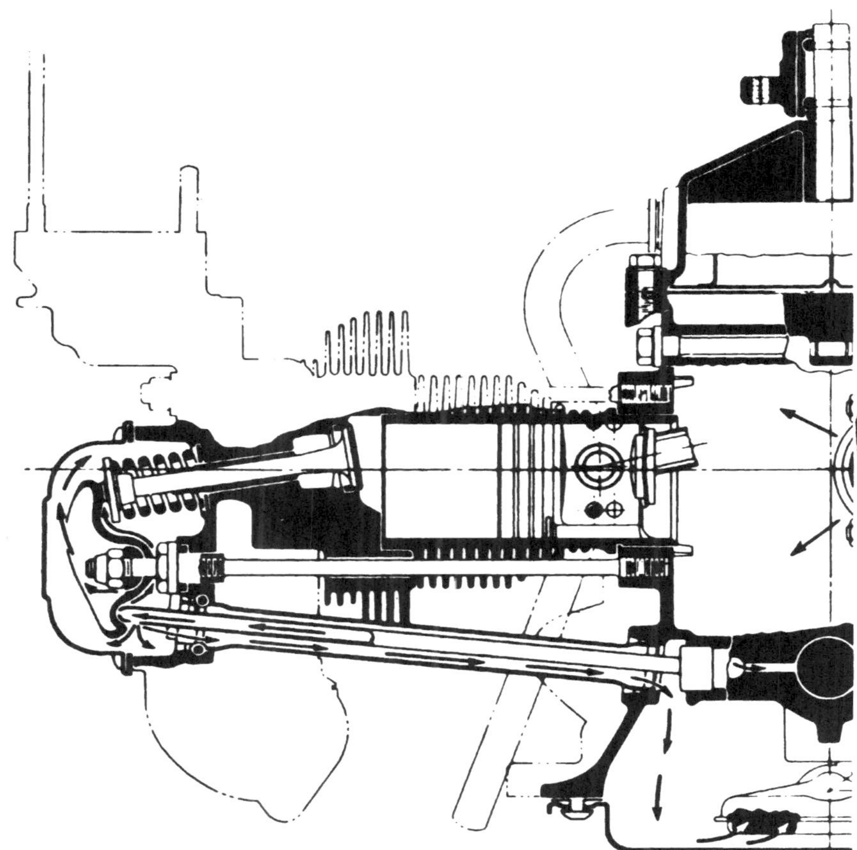

Lubrication-system changes are essential for engine life when building a racing engine. This chapter tells why the changes are needed and how to make them.

Chevrolet improved fan-belt life by changing to a magnesium blower fan (1964), but the lesser inertia of this new component didn't help belt-loss problems very much. The added belt guides did help. Information is provided on installing the magnesium blower and perhaps more importantly, minor idler-pulley improvements to ensure belt retention. Details of the pros and cons of changing blower speed through use of special pulleys are additional features of this chapter.

No matter what you do to help the fan belt installation and cooling performance, you need to pay attention to the belt. No matter how much care you take in pulley alignment, belt lubrication and maintenance, you will eventually throw a belt. The first indication you will see is the temp-press light coming on as the alternator is no longer charging. If you are "enthusiastically" driving the car, you will probably feel the few extra horsepower that you have at your foot's disposal, since you are not spinning the fan. There is a new indicator kit available that will tell you almost immediately if the belt comes off and cooling air pressure has dropped off above the motor. Levair Performance is selling a kit that reads the pressure in the upper shroud and turns on a warning light if the pressure drops below a safe threshold. If the belt comes off at an inopportune time, you might only have a minute until the high head temps convince the valve seats to leave.

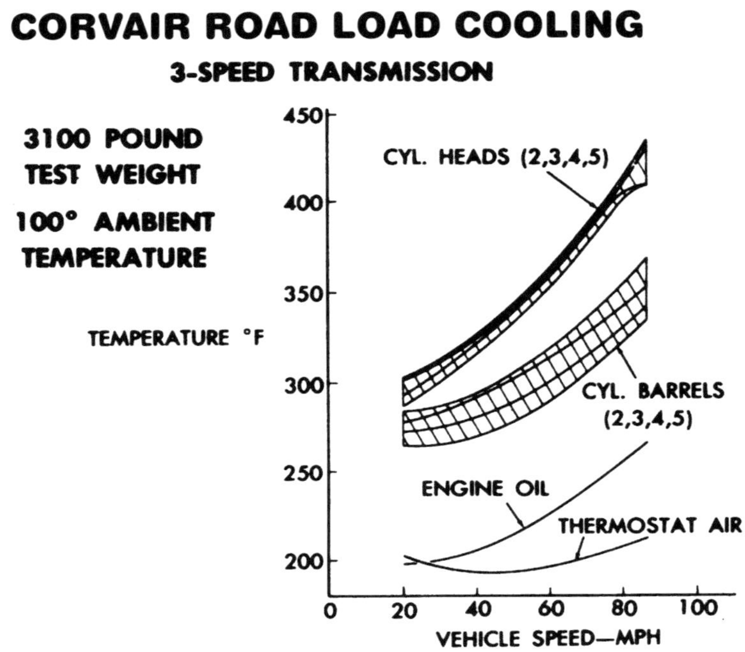

Stock Cooling System

Although it might appear that the Corvair is the same yesterday, today, and tomorrow, nothing could be farther from the truth regarding its cooling system. Detailed changes that have been made deserve your careful consideration and understanding.

The Corvair uses forced-air cooling. Engine enclosure by sheet metal shrouding directs air over the cylinder heads, cylinders, and oil cooler. Thermostatic cooling control is provided on 1961 and later engines by bellows-controlled damper doors at each side of the engine. These doors open when temperature of the air under the cylinders attains 205° F. Thus, even on cold days, the engine is operated as if it were in an ambient temperature of approximately 80° F. An engine blower fan is mounted horizontally atop the engine. It is belt driven at 1.58 times engine speed. 1960 models differ in that cooling air is controlled by a thermostatically controlled ring on the blower-fan inlet.

How Hot Is the Engine Getting?

The factory keeps you in the dark on this question, at least when you are driving the car. Nothing warns you of an overheat condition except an instrument-panel light which illuminates in the event of low oil pressure or an overheat condition sensed at one cylinder head. Oil temperature is neither measured nor indicated. Although cylinder-head overheat is indicated immediately by the light, oil pressure indication is far too slow. By the time the light comes on the damage has probably already occurred. For this reason an oil-pressure gauge should be considered even more important than the cylinder-head and oil-temperature gauges or warnings which are detailed herein. It should be mentioned that the Spyder head temperature circuit includes a buzzer to "sound off" and provide an insistent indication that it is time to "back off." This feature was carried on for the Corsa model, when turbo equipped. On non-turbos, the wiring is still there, they just left the buzzer out. The Temp-Press indicator lights up simultaneously on the dash.

The importance of knowing more about temperatures in a modified engine is stressed by the following curves and description for the original SAE Paper 140C, "The Chevrolet Corvair."

"An air-cooled engine is operated over a considerable temperature range under normal operation. From 20 miles an hour at road-load conditions to maximum speed, the cylinder-head temperature increases from about 300 to 435 degrees. This represents the worst road-load cooling condition: the same temperature also prevails at lower speeds during full-throttle operation. The temperature characteristic curve of this engine follows quite closely after the indicated mean effective pressure curve. The peak for our engine, occurring at about 2,600 RPM represents the worst cooling conditions we can obtain—about 30 degrees hotter than at maximum speed."

The foregoing statements related to the 80 HP engine as originally introduced and failed to mention that oil temperature simultaneously soared to 280° F, which is higher than can be considered safe. According to Mackerle, oil temperatures should be maintained above 176° F to keep down friction loss and dilution by gasoline, and should not exceed 230° F for continuous operation. Up to 230° F may be considered permissible for short bursts.

Measuring Cylinder Head Temperatures

All Spyders and the 1965–66 Corsa models have cylinder head temperature gauges operating from one cylinder head. Like many low-production parts, these devices are quite high priced should you choose to build your own from the stock parts, if they are available. You could install just the one gauge, but it would require a separate housing, and the marked lens must be mounted in the housing that you fabricate. The harder part relates to removal of your standard cylinder heads and machining an opening for the sender—that must match the one used in the Turbo 140 head, so you'll need one of those for a sample. Several sources sell an adapter which lets you install the fine-threaded sender in place of the original coarse-threaded unit. Perhaps a simpler method would be to change to an aftermarket Cylinder Head Temp (CHT) gauge. They used to be commonly available from most gauge manufacturers, but since Porsche left the "air-cooled" scene, many of them have been dropped. VDO still sells an appropriate gauge. Westach Instruments in Petaluma, California, makes both a single and a dual CHT gauge. Autometer still sells one, as long as your CHT will never exceed 340° F. The GM gauge was a "thermistor" based gauge, where the sending unit changed its resistance to ground as the temperature rose. All of the aftermarket gauges are "thermocouple" based, where a pair of wires of different metallurgy, crimped together,

Westberg Manufacturing makes Westach gauges, including this dual cylinder head temp unit.

This single head temp gauge makes a great replacement for the factory unit on the Spyder and Corsa.

will actually send a small current down the wire pair, when the junction is subjected to the change in heat. The gauge on the dash is calibrated to show that small current as a temperature reading. The aftermarket gauges come with a method of installing the thermocouple into or onto the head. Many of them use a ring under the sparkplug to carry the thermocouple. The most accurate ones use a thermocouple probe which can be placed in the original thermistor locations, up inside the cylinder head. Be sure that the gauge will display a range of temperatures up to 600° F.

A Chevy dealer bulletin advised Spyder cylinder-head temperatures as: 200° to 300° at idle; 350° to 475° at 30 to 60 MPH cruise; and 460° to 575° from 3,000 to 5,000 RPM at full throttle. Overheat-warning switches are used on all Corvair engines, and these have changed as the factory has upped the horsepower produced by the engine. The unit used on the Spyder and 140 HP engines operates at 575° F, but uses a 3/8 inch-24 thread instead of the 3/8 inch-16 thread used on the other switches. '60–'64 engines, if not originally supplied with air conditioning, can have a switch that operates somewhere close to this 575° F temperature by installing Part No. 1993574 (identified by grooves on the hex nut).

Oil Temperature Measurement

Oil temperature can be measured with a sending unit coupled to a gauge. Although all special oil pans include a boss for the addition of such a sender, the bottom of the oil sump gives a false (too-cool or too-hot) reading that cannot be relied on for accuracy. The best place to measure oil temperature in the Corvair—as the oil leaves the cooler—is not easy to get to because the rear accessory cover delivers the cooled oil directly to the main oil passages. A second-best choice for early engines is to insert the temperature sender in the right side of the block where the pressure sender was originally located. If this spot is used, an aluminum heat shield must be constructed for the sender. Later blocks have a boss near the cooler. This boss can be drilled and tapped for sender installation.

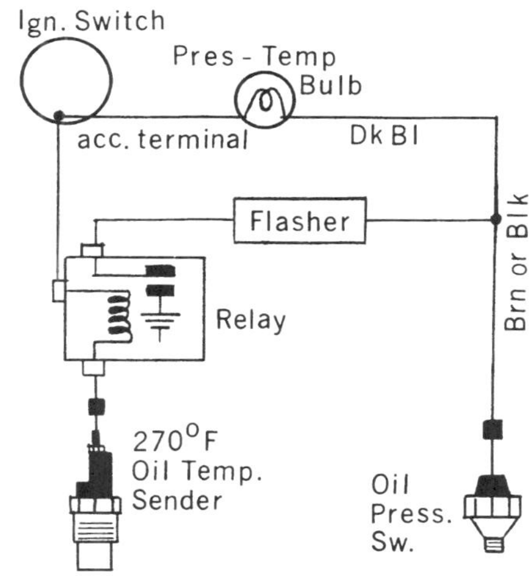

Ford oil-temperature switch closes at 280° F, flashing panel light in warning when wired as shown.

If you merely want to be warned when oil temperature reaches the danger point, some small amount of money will buy the system Ford uses with their 427s. The schematic diagram shows how to wire the parts to work with your TEMP-PRESS warning light. When oil temperature rises above 270°+/– 48° F, the instrument-panel warning light flashes on and off until temperature drops below that point. The low-pressure warning (2 to 6 psi) operates as it was intended to. Ford parts required are: oil temperature relay

Autometer makes several different series of instruments, many of them resemble the original GM-style gauges.

ClTFlOB840-A, oil-temperature flasher C3AZ9E296-A, and oil-temperature sending unit C3AZlOB921-A. The sender will require some special mounting techniques.

Pressure Switch

When your Corvair engine's oil pressure drops to 2 to 6 psi, the oil-pressure warning light tells you that you have insufficient oil pressure. Racers sometimes call the original switch, the "Engine Destruct Indicator," because it comes on at such a low pressure, telling you why the motor just broke—you did not have enough oil pressure. This inadequate warning system is the primary reason for installing an oil-pressure gauge. 1960–62 pressure switches failed and leaked frequently due to design and location in an area of extreme heat in the air-exhaust duct. The switch was improved and relocated. Modify your '60–'61 model to relocate the new switch into the top of the oil-filter adapter. Use a six-point 1-1⁄16 socket, not pliers or a 12-point socket or you will distort the switch, causing quick failure.

Oil Pressure Gauge Choices–Electric versus Mechanical

Installation of an electrical-type gauge is comparatively easy, as it is only necessary to mount the stock oil-pressure switch with the pressure sender onto a tee. It is important to keep the oil pressure switch and warning light in the system. You might just be looking the other way when the oil filter bolt falls out. The light will draw your attention. It is safer, and therefore more desirable, to measure the pressure at a point more remote from the pump output, such as at the main oil gallery. The tapped hole on the right side of the block where the oil-level gauge tube enters the block connects directly into one main oil gallery. Measuring pressure nearer the pump outlet may not disclose a blocked oil gallery, but that is pretty rare. Should you choose to use the remote location for the pressure-measuring point, then the sender should not be mounted directly onto the block below the cylinders.

The first Corvairs ('60 and '61) suffered from pressure switch failure due to high ambient temperatures and this location was abandoned in favor of placing the switch by the generator. A 45-degree elbow can be screwed into the boss and a copper tube used to carry the oil to the pressure switch that can be bracket supported just atop the center of cylinders 1 and 3. Take care not to obstruct the airflow, of course. Mechanical pressure gauges, even with the long pressure tubes needed when locating the actual gauge at the dash, used to be preferred to the mechanical gauges, because the electrical ones were limited to a 90-degree movement. The mechanical pointer could sweep about 270 degrees, giving finer definition of the temps. Newer electric gauges, now with internal stepper motors, can sweep the larger area.

Gauge Installations

If you are preparing an early model that didn't start out as a Spyder, you can switch to the Spyder dash and the instrument feeds to provide the data. That will present you with a few extra gauges. But you will still be missing an oil pressure and oil temp gauge. A small panel underneath the radio will provide you with room for three extra gauges. I suggest an oil pressure, an oil temp, and a voltmeter. Late models can add gauges in the dash above the radio, even angling them toward the driver. But if you start with the Corsa dash, you can add several gauges in the dash itself, without sawing new holes or adding mounting plates underneath, maintaining that "stock" look. If you are not running a turbo, you can dump the vacuum gauge along with the small clock that hasn't worked since the Carter administration. You can replace the lame Corsa tachometer and even install a GPS enabled speedometer if you wish. Now on the market is a set of four mounting plates that attach to the back side of the Corsa dash, using the original mounting holes. The plates retain the standard aftermarket gauges of your choice, four in the middle, 2-1⁄16" gauges, and two on the side, 3-3⁄8" gauges. The late Corvair used the standard 0-90 ohm fuel level sender and all the gauge manufacturers make compatible gauges. The plates are removable and the installation can be returned to stock if you change your mind. See the pictures and start picking out your gauges. You need not replace the Corsa speedometer, if it is working okay, unless you spend a lot of time over 140 MPH.

Corvair Engine Blowers (Fans)

Four types of Corvair fans have been used during 1960–64. All are described here and three are pictured.

1960–mid-1961—11-inch diameter steel fan with 24 vanes on an outer ring. 1,800 CFM at 4,000 engine RPM.

This view of the back side of a Corsa dash shows how individual gauges can be replaced with aftermarket gauges, including those important oil pressure and temperature gauges. Everything screws onto the original mounting points and is totally reversible.

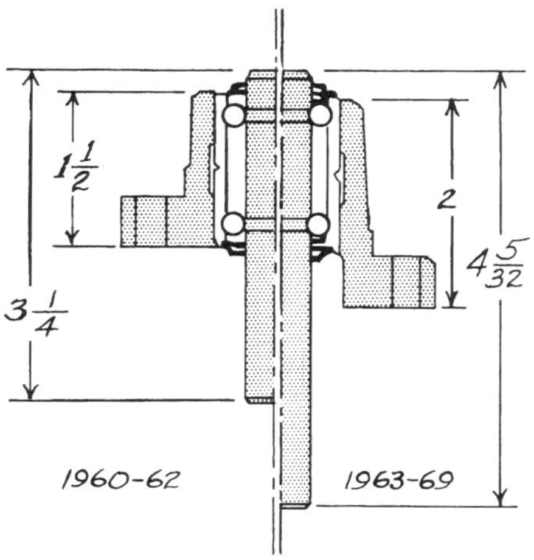

Blower-hub comparison. Later models place load near center of double-row bearing for increased life.

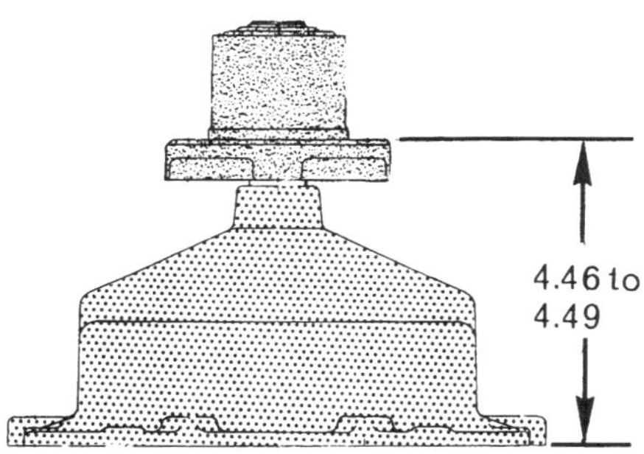

Installation drawing shows height required for installing late hub and shaft assembly into the crankcase cover.

Poor air-delivery characteristics at high RPM. Ring on inlet probably reduced theoretical output.

1961–1963—Production-line "running change" made in mid-1961 to a 10.7-inch diameter steel fan with 16 vanes extending to the center of the blower hub. 1,460 CFM at 4,000 engine RPM.

1962 Plastic Fan—1962 engines were supposed to be equipped with a plastic fan (nylon) and GM literature indicated this fact. Few, if any, got into the hands of private owners as these were withdrawn from production and replaced by dealers at factory request.

1964–1969—11.2 inch diameter magnesium blower fan. Design almost identical to plastic fan that was abandoned. Same air delivery as 1961–63 models, but weighs ⅓ as much. Provides greatly improved belt following at high RPM. Belt is less subject to stretch or slip on acceleration and deceleration because there is less fan mass. Installation of this blower requires minor blower bearing and hub changes for best bearing life. The '64 unit can be installed directly onto the 1960–63 hubs without change if desired.

NOTE: *CFM = Cubic Feet per Minute*

How To Install the 1964 Blower (Magnesium) on Earlier Models

Remove the main shroud and the old blower. Install the 1964 blower fan directly onto the hub. The magnesium fan, Part No. 3828742 sold for $6.40 new, and are commonly found at swap meets. The '65 and later fan used different bolts and isn't a direct bolt-on for the '60–'64. If the engine is to be used in an all-out-racing situation, 1/2 inch can be turned off the major diameter of the magnesium blower, according to an article in Corvair Communiqués, Vol. 2, No. 3.

While you have the shroud and old blower off, you should first check the bearing condition prior to installing the new fan. If the bearing is noisy or gritty feeling, it should be replaced. An improved bearing and hub assembly

Another American Pi production. The fan bearing assembly comes apart to replace the bearings.

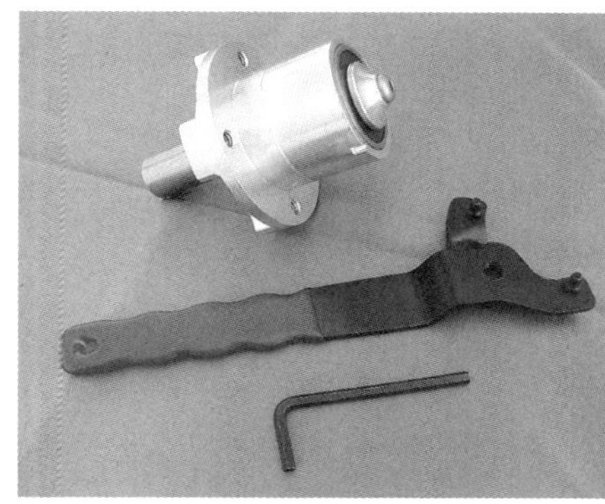

As assembled. The special tool holds the fan bearing in place as you remove the top retainer.

was introduced on 1963 models on engines beginning with TO 913. This design placed the blower-pulley centerline ⅝-inch closer to the bearing centerline, thereby improving life through a better load distribution. Parts required are an Engine Blower Pulley Hub, 3826474 and a Blower Bearing & Shaft Assembly, 907448. Note that both hub and bearing are designed as a pair and must be installed together. Remove the crankcase cover. Support crankcase cover and press blower shaft out. Remove blower bearing and shaft assembly from hub by pressing on bearing outer race with a ⅞-inch socket (deep type). Install new bearing and shaft to hub.

WARNING: *Press only on bearing outer race when removing or installing bearing and shaft assembly to blower hub. Do not press on inner race or shield.*

Coat shaft with hypoid lubricant or Lubriplate. While supporting crankcase cover, press on shaft to install hub/bearing assembly to height shown in drawing.

Before replacing the crankcase cover you will want to read the section on crankcase breathing to see what further changes or additions may be needed prior to reassembly. Also read details on special pulleys.

Fan Belt

The best data on fan belts was published in Chevrolet Service News in the January 1962 issue. “Since the introduction of the Corvair vehicle line, exhaustive laboratory and field testing has been conducted to develop a Corvair engine blower belt that will provide maximum durability. As a result of this constant testing and improvement, it is felt that the currently used blower belt #3780981 will provide very satisfactory service . . . the only fan belt now known which will meet Chevrolet engineering specifications. Corvair owners should be told that if in an emergency, it is necessary to install a different belt, it would be advisable to replace that belt with #3780981, as soon as convenient.” Of course, Chevy hasn’t supplied that belt in a long time. Otto parts supplied a fine belt, the Otto Parts “Super Belt.”

FAN POWER TABLE

RPM	*Stock Pulley*	*Adjustable Eelco Pulley*			
	1.58:1 ratio	*1.4*	*1.32*	*1.25*	*1.20*
2,000	1.0	0.9	0.8	0.6	0.5
3,000	3.5	2.6	2.5	2.0	1.9
4,000	8.0	7.0	5.2	4.0	3.7
5,000	15.6	10.6	8.3	8.0	7.5
6,000	27.0	18.5	16.5	13.4	11.8
	Numbers in these 5 columns are HP required by fan at RPM shown				

Eelco’s adjustable pulley, now available from Clark’s, provides four fan ratios to increase HP. Installation also requires use of a longer belt.

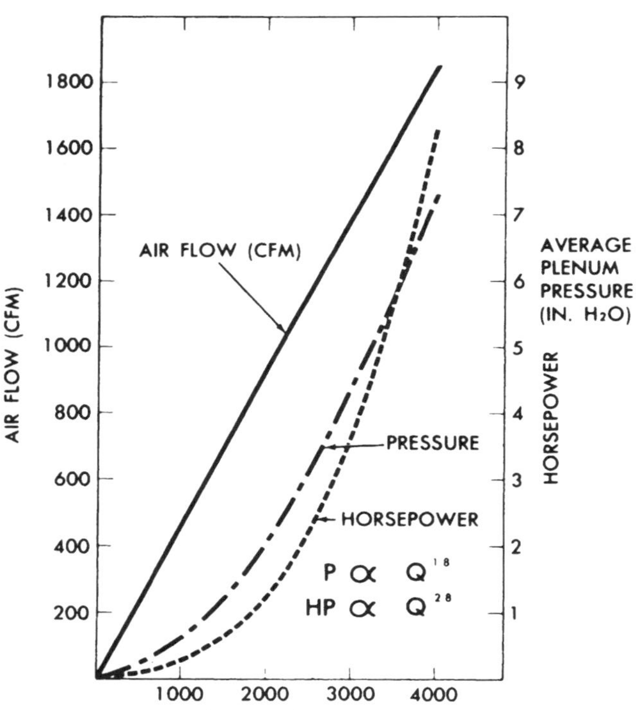

BLOWER POWER REQUIREMENTS

Drawings adapted from SAE 140c.

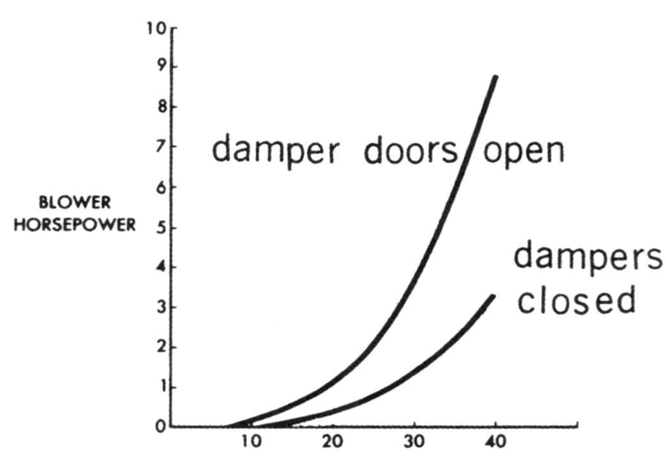

That belt is still supplied by Clark's Corvair. Clark's also offers their own HD belts. I have found the narrow belts that ride farther down in the pulleys have better life. The Gates 3VX560 belt has also worked well. If you ask 10 Corvair racers, you will get 10 answers, maybe 12.

"Most owners carry a spare belt for emergencies. If a belt flips over, replace it. Some of its strands probably parted when it assumed the reverse position." If you are going to store a belt for emergencies, don't coil it up too tight, as it will have to be stretched to operate.

Keeping the Fan Belt Intact

In mid-1964, Chevy introduced simple belt guides to restrain the belt from flipping over or leaving the pulleys at high RPM. These guides have proved especially helpful, both for normal and competition driving. However, they do have sharp spots on them as they come from the Chevrolet store, and it is a good idea to round off all of the sharp edges. The rear guide installs easily on the idler pulley mounting studs, but the front guide requires a homemade wedge-shaped shim to make the installation on shrouds supplied prior to the introduction of the guides. It is easiest today, to just swap out the top shroud.

Don Eichstadt points out that fan-belt throwing can be caused by crank pulley or balancer run out. He suggests six radial paint marks around the balancer rim so that creeping of the outer rim can be noticed easily, no matter what position the balancer is in when the engine is stopped.

Fan-belt life can sometimes be improved with a spring-loaded idler assembly, but this is not a cure-all. Careful attention to the alignment of all pulleys and the use of stock belt guides will usually keep the belt intact and on the job. The spring-loaded kits were made by EMPI, OTTO Parts, Bill Thomas, and Corvair Underground.

Installation of the spring-loaded idler is extremely simple, especially with some of the kits that are available. A bracket is screwed to the sheet metal at the rear of the engine compartment. The idler is removed and a hole drilled for spring attachment. The Otto parts kit used a twisted metal rod for attachment. A ½-inch drill is used to enlarge the existing hole in the idler bracket. An oversize bolt is used as a pivot point. The forward stud is equipped with a large flat washer adjusted with a lock nut. Both sides of the idler are coated with high-temperature Lubriplate to allow the idler to pivot freely in the fore and aft directions. Details for installation are included in the kits.

Spring-loaded idler is easily installed, acts as a shock absorber to keep belt on pulleys and to increase belt life. 1964–65 magnesium fan recommended. Otto Parts idler illustrated.

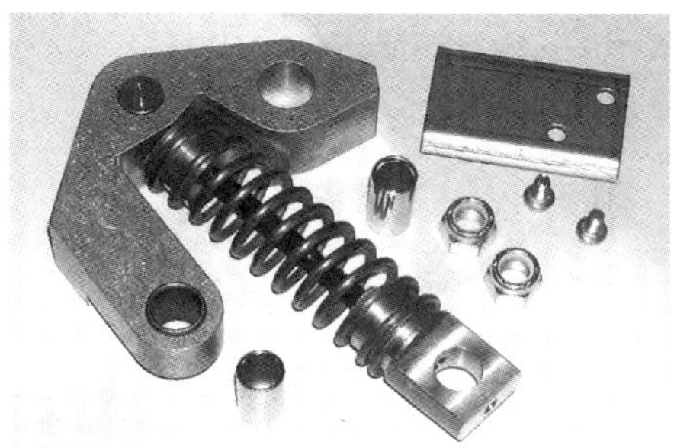

Spring idler kit as received.

Corvair Underground fan belt idler kit with pulley installed.

This aluminum fan drive pulley from American PI replaces the stock, stamped-steel unit.

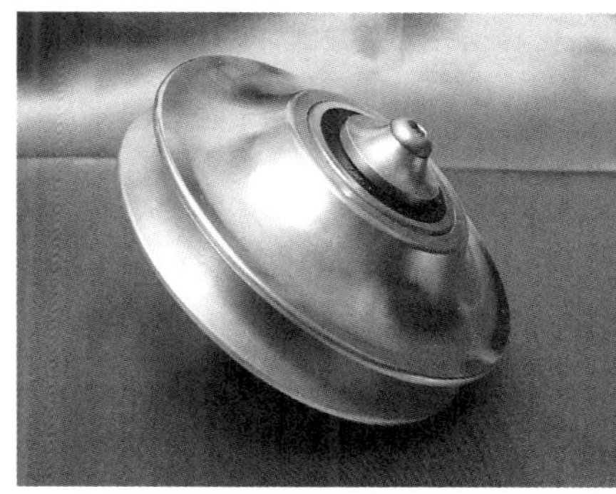

This aluminum idler pulley uses readily available bearings and is machined to a tight tolerance.

The Corvair alternator rotates the opposite direction from the V8 units. This Corvair-specific fan unit will pump more air than the stock unit and look great as well.

This is the fully assembled unit.

After installing the belt on the crank, fan, and idler pulleys, the spring is attached. Then, while holding the idler in the forward position, the belt is slipped onto the generator pulley.

Incidentally, the most-important alignment is that of the generator pulley, which must be positioned about ¼-inch rearward of what would appear to be an optimum straight line connecting with the crankshaft pulley. GM tests showed that this eased the belt onto the larger crankshaft pulley and prevented the tensioned belt from flipping over. The same tests showed that alignment of the belt leaving the crank pulley had the least effect on belt life.

The newest kit on the market is the one from Corvair Underground. The spring-loaded portion of the idler is built into the pulley. No need to attach a spring to the rear body work. The unit has an adjustable spring assembly so you can adjust the amount of tension applied to the belt. This has a lock nut which must be tightened up before running the unit. The mounting on the engine is about the same as the older kits. The trick is providing a restrained mount that allows easy travel fore and aft for belt tension, but almost no side to side movement which would affect the fan drive pulley and the dampener groove alignment. Finally, for the racers, a new complete kit is on the market, the half-speed fan kit from LeVair Performance Products. This kit has you replace a number of different parts on the motor. The fan drive pulley on top is larger, the vibration dampener is replaced by a modified dampener inner pulley, with the outer ring removed and a smaller-than-stock driving pulley installed, and finally, the upper oil-filter/generator adapter is modified for special alignment. With the smaller driving pulley and the larger driven pulley, the actual speed of the fan is much slower than stock at any given engine RPM. But with those two modified pulleys, the plane that the alternator rotates in has changed and the plane that the spring loaded idler pulley rotates in also has to change. The upper adapter housing determines those planes because both items attach to the sides of the adapter. The LeVair modified adapter has both sides remachined to better align the pulleys with the relocated drive and driven pulleys. It is slick and it works. It also uses the regular Corvair fan belt. It is for racing/autocross use only, however, because even with the full size stock fan, not enough air is moved at slow speeds to adequately cool the motor. It is fine at high RPM, but not for street use. I use this kit with the Corvair Underground pulley on my autocross/track car.

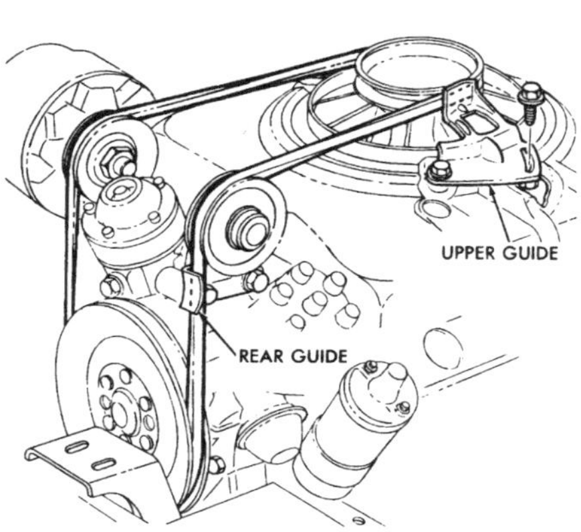

Installation of 1964-65 belt retainers requires using ⅟₁₆-inch shim between belt and retainer by idler, then tensioning belt and tightening idler. Shim is next used to set clearance by fan pulley. Illustration from 1965 Corvair Shop Manual.

1) Smaller drive pulley replaces harmonic dampener. 2) Larger driven pulley results in slower fan rotation. 3) Alternator adapter is modified to align the belt guides to the altered pulleys. (Oil filter was moved to the front of the car.)

Why Special Fan Pulleys?

Adequate cooling for the worst load and ambient-temperature conditions, including full-throttle operation in summer temperatures is provided by the stock system. Turning at 1.58 times engine speed, the fan produces 1,460 CFM at 4,000 RPM (engine speed). This air quantity is just sufficient for continuous full-throttle operation and is absolutely necessary for hard driving in hot climates or for road racing.

Horsepower is needed to drive the Corvair fan and the same is true for a water-cooled engine's fan and water pump. But, the water-cooled job has the additional task of pushing air through the radiator. This means no streamlined airflow and increased wind resistance. Thus, when all factors are considered, the air-cooled engine probably requires less HP for cooling especially at part-throttle cruising. Look at the graphs and notice that only 3.5 to 5.5 HP is required at 60 to 70 MPH in high gear. Fan-power requirement increases as the cube of speed and is dramatically detailed for stock and special-ratio pulleys in the Fan Power Table, page 88.

For street or highway operation where ambient temperatures are not extreme, reducing fan speed adds horsepower that you can feel! When a special pulley is installed you must become alert to the overheat indicator because less cooling is available at a given speed. And, you may want to install oil-temperature warning devices as described elsewhere in this chapter.

To install just a special pulley requires using a longer belt and realigning generator and idler pulleys. Because the longer belt runs at the same speed (crankshaft-pulley diameter has not been altered), belt wear stays the same. But, because the longer belts are not specifically engineered for the Corvair's tortuously twisted layout, life may not be equal to the stock component. In any event, the use of a spring-loaded idler is recommended for best belt life.

One of the slickest Corvair accessories was Eelco's adjustable fan pulley, also sold by Otto Parts, still available at Clark's. The pulley is made in two pieces that are held together by four bolts that also attach the device to the blower hub. Construction is such that the parts can be rotated in relation to each other, then bolted solid, to obtain four separate ratios as shown in the preceding table. The 1.2:1 groove is at stock groove height, but the lower ratios (smaller diameters) move the groove upward. This places the 1.4 ratio ¼-inch too high for perfect alignment. To make best use of these lower ratios the blower-bearing shaft can be pressed farther into the crankcase cover to place the groove at stock height. I recommend that 1.32:1 be the slowest ratio used so as to maintain adequate groove depth.

If a single-ratio pulley is used, the blower-bearing shaft can be shifted to keep the groove at stock height.

Damper Doors

One of the bits of misinformation that prompted this book was the widely believed fiction that the Corvair's thermostatically controlled cooling system should be disabled. One performance manual went so far as to recommend that the damper doors should be propped open permanently or the inlet-damper ring eliminated on the 1960 models. This same writer further stated that "no damage could result and this would reduce the amount of horsepower required by the blower fan." Thus, many Corvair owners have obtained two misconceptions.

Let us examine these "recommendations" in detail. First, blocking the dampers open increases fan-horsepower requirements. When an increased volume of air is moved, as when the dampers fully open for maximum cooling, it takes more power to push the air through the cylinders and the oil cooler. Fan HP requirements are reduced by closing the dampers and the recirculation ducts. For drag racing you can block the dampers closed for the race, then open them immediately afterward. If a working generator is not required by the rules, then remove the belt for the race and simplify the entire procedure. This assumes that you have installed the spring-loaded idler to make this a convenient procedure. Secondly, blocking the dampers open increases engine wear by lengthening the warm-up time. Open dampers eliminate one of Corvair's most

Brian O'Neill's track car uses an electric motor drive for a full-size fan. Proven in cool weather, it has yet to be tested in hot conditions.

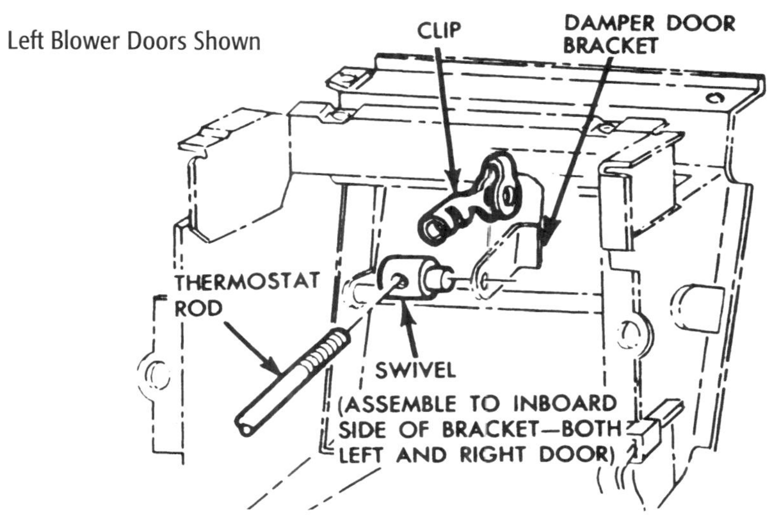

Early model shroud and thermostat assembly.

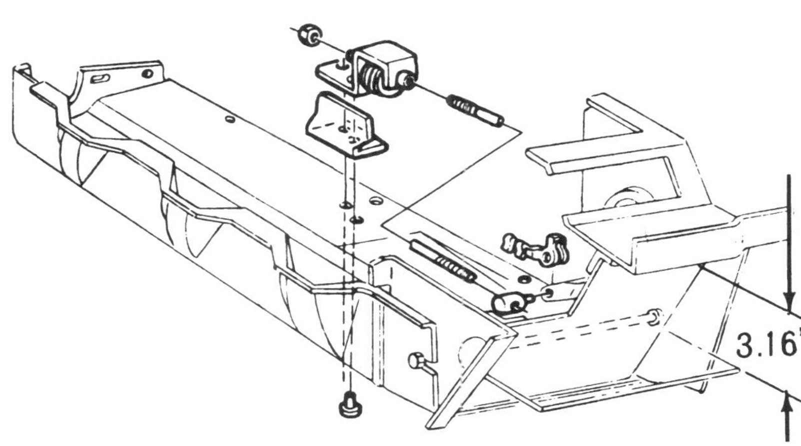

Late model drawing shows that lower shroud and thermostats were revised in '65, placing the thermostat just behind the center cylinders and shortening the operating rod.

desirable features—quick warm-ups. A quickly warmed up engine suffers less wear, especially when it is allowed to reach and maintain its normal operating temperature for a period of time prior to shutting it off.

Short hops during which an engine never warms up increase crankcase dilution by gasoline, build up varnish and sludge accumulations, and greatly increase cylinder wear. The high-wear rate is caused by combustion products that condense on the cylinder walls to cause etching and rapid wear. Cold cylinders also mean thick oil that increases the friction that pistons must overcome. The hotter the cylinder walls, the less friction loss, especially with high-viscosity oils. This is another good argument for the use of multi-viscosity oils: to decrease friction during the critical warm-up period. If you are operating the Corvair engine in an extreme environment for extended period—over 100° F and high loads, you could remove the lower shrouds for those conditions. Put them back on when the temps drop back to normal.

You can check this oft-proved and well-documented series of facts in any internal combustion engine textbook.

Corvair Damper Door Assembly

When adjusting thermostat linkage the following detail must be observed. The thermostat-rod swivel, at both the left and right damper door, must always be assembled to the inboard side of the door bracket. The swivel pin must always point outboard as it is inserted into the door bracket. Improper assembly will cause a bind at the thermostat stem, resulting in leakage and failure of the thermostat when the engine is operated. There are several tricks to connecting up the link to the doors. It is important that you do not let the doors snap closed during installation or removal of the shrouds. That hard snap closing will damage the thermostat in short order, and good ones are getting expensive. I assembled a special tool for connecting the thermostats. It can usually be used for removal as well. I use an old thermostat rod with a long fine thread nut installed and double-nutted into place, halfway onto the rod. I pull the door open and screw the other end of the double-long nut onto the tip of the installed rod. You can now pull gently on the rod to take the pressure off the thermostat and allow the door to move freely. A screwdriver tip can lift up the "pal" portion of the clip and you can just slide the swivel pin out and the clip off. It works for the installation as well.

Oil Coolers

Like the cooling fans, the oil coolers have also been subject to considerable change since the introduction of the Corvair engine in its original form. At least four coolers have been supplied during the 1960–67 period. The original folded-fin aluminum oil cooler as used on 1960–62 models is the best one you can get. It dissipated about 160 BTUs per minute, keeping lubricating oil temperature at 280° F with the engine under full-throttle operation in an ambient temperature of 100° F. In 1963, the production-economy types introduced a new design 3-plate cooler on standard and 102 HP engines, and an 8-plate job on the Spyders. Experience showed that more cooling was needed for the 164-inch engines introduced in '64, so those used the 8-plate design on standard and 110 HP engines, and a 12-plate on the Spyders. The '65 Corsa and 140 HP engines all use the 12-plate models, lower HP models got the 8-plate. When both are perfectly clean, the 12-plate type doesn't shed as much heat as the original folded fin cooler, according to all of the information that I have been able to glean from various sources.

If you are going to race your Corvair in hill climbs or road races, then improved cooling must be supplied by installing a large, remote-mounted cooler that is connected through ½-inch inside diameter (AN-8) or larger hoses. The hoses can be connected to the stock oil cooler adapter manifold after it has been modified by heliarcing female pipe fittings (aluminum) to it and opening up the holes. The reason for the large lines is that oil is quite viscous when cold and large lines ensure that the bearings will be adequately supplied under warm-up conditions. The pressure-relief valve in the filter adapter must be removed and plugged. Otherwise, the restriction caused by the cooler and filter combination will cause the relief valve to open and the oil will bypass the cooler. Before modifying the oiling system in any way study page 6A-6 of the '61 Corvair Shop Manual where lubrication system functioning and routing is thoroughly explained and diagrammed.

Three of the four oil coolers supplied for Corvairs from 1960-2 (top), 1963 Spyder and '64–'69 standard and high-performance (center), and (bottom) is the 12-plate cooler for '64 Spyders, '65 140–180 HP engines, and all air-conditioned engines.

Doug Roe's Spyder-engined "Desert Rat" never exceeded 260° F oil temperature in a one-hour race in 90° ambient temperature using Harrison "big car" oil cooler installed as you see it here. Body is cut away behind cooler for airflow. Hose is cool-air inlet for carburetor.

The best externally mounted oil cooler is the stock car racer's model that Harrison Division of GM manufactured. Because that is no longer available, the aftermarket has stepped up with dozens of different coolers available in every size and shape. The mounting shown in Doug Roe's car is generally accepted as the best position as cool air can be ducted to the cooler, and it is close to the engine so that the lines can be kept short. Most race organizations now allow the cooler to be mounted up front, usually behind the license plate area, the final Stinger location. If you are ducting air to the oil cooler, it pays dividends to duct air away on the outlet side.

Oil Filters

The stock Corvair oil filter is adequate for most road use, but when the engine is being modified for competition, use a larger filter. It will give added oil capacity as well as improved filtering. A non-stock filter will also be easier to mount remotely on Porsche/VW installations where the stock filter has to be cut off to get the engine into the chassis. Perhaps the easiest full-flow filters to adapt are those made by Ford. The racing aftermarket has embraced this design and produced many adapters to allow remote mounting of single or dual filters. The Ford filter, and the aftermarket equal, contains a bypass valve inside each filter, unlike the Chevy V8 filters which have the bypass mounted in the block. The remote adapters start at about $15. Trans-Dapt P/N 1028. Any of the better Ford replacement filters will work. I recommend the K&N spin-on HP-3001. It is a Ford replacement and is a very well-made part. There are also smaller filters that fit, if you have a space problem.

Hoses to and from the oil filter must be ½-inch diameter minimum, I use AN-10 hose. These inside diameters are absolutely essential to avoid oil starvation when the oil is cold and viscous. Attaching the hoses at the engine end is another problem, especially with the die cast oil filter adapters used after 1961. Adding threaded holes to attach ½-inch I.D. hoses will probably require heliarcing onto one of the 1960–61 permanent-molded adapters as these have more "meat" and are more easily modified. Do not hesitate to reroute the passages in the adapter to permit getting the hoses in the correct position. You may want to bring the new hoses in under the generator, or into the top of the adapter, depending on the rest of the installation problems. Avoid interference with the fan belt and generator. Arrange the hoses so that they do not have to be removed to install a new fan belt.

Oil Pans: Stock or Special?

Finned sump covers should only be discussed after understanding the engineering facts of the matter. Fortunately, several authorities on engine construction have written a considerable amount of material that we can study to improve our knowledge. Besides his motorcycle design experience (Vincent motorcycles) Mr. Irving was the designer behind the F1 Champion Repco Brabham V8 engine.

P. E. Irving, in his excellent book *Motorcycle Engineering* has this to say, (p. 239):

"While oil is good at collecting heat, it is very bad at getting rid of it again, because the layer directly in contact with a cool surface increases in viscosity and simply stays there, acting as an insulator and effectively preventing heat being dissipated from the hotter oil in the interior. Ribbing a sump which contains a quantity of oil is not very effective unless there are some internal ribs also to transfer as much heat as possible from the body of the oil, but ribs placed on areas against which hot oil is violently thrown by centrifugal action can be made to radiate a lot of heat. In this connection, the polishing of crankcases, though pleasing to the eye, may cost almost as much as the whole of the machining and cuts down the heat-radiating ability to a fraction of what it would be if the metal were left "as cast."

In another part of the same book (p. 183) Mr. Irving says,

". . . a polished surface emits less heat by radiation than a black one. The rate of emission from a polished surface is approximately one-tenth that from the same surface covered with a thin film of lamp-black, and the emissivity of a cast-aluminum surface is increased about 10 per cent by a *thin* coating of black paint."

Note that a black surface is ten times as efficient as a polished one. The Alcoa Engineering Handbook indicates that Irving's comment may be quite conservative. Alcoa compares an as-cast surface with one which has been black-anodized to a depth of 1.7 thousandths. The black surface is more than ten times better in heat-radiating ability than a plain cast surface. Remember these facts when you are tempted to start polishing and chroming various engine parts that could conceivably contribute to cooling efficiency.

Another noted mechanical engineer, Mr. Julius Mackerle in his book *Air-Cooled Motor Engines* states that it is an error to assume that using a greater quantity of oil will reduce oil temperature. He further remarks that, "A finned sump does not aid cooling to any great extent as the oil does not flow down the cooled sump walls. Cooling is more intense on the crankcase walls, over which the oil flows in a thin film...Best oil cooling is obtained by a tube-type radiator . . . "

The reader should take note of the words "thin film" as these are the key to understanding the removal of heat from oil. If you choose to use the stock Corvair oil pan, do not chrome plate it! Additional heat-removal capability can be added by welding or brazing sheet metal fingers and/or baffles to extend into the hot oil to transmit the heat to the radiating surface. The pan is black as received from the factory. If you have to refinish it, use a thin coat of flat-black paint. The stock pan has approximately 178 square inches of radiating surface.

Should you decide on one of the more exotic special pans made of cast aluminum you will find that there are several to choose from. Most of them offer similar features including:

1. Baffles, which slow oil in sloshing away from the pump pickup in corners or on hard acceleration and deceleration. These baffles serve an important

The newest Clark's pan is finned on the bottom side to maximize temperature shedding.

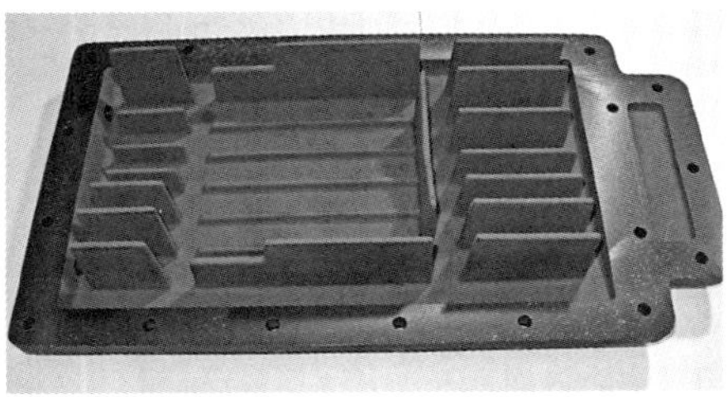

Inside the pan, these fins gather heat from the oil, by maximizing contact area.

function in conducting heat to the radiating surface of the pan, providing the pan is not polished. Some pans have no baffles, or very short baffles that should be lengthened to at least provide one central baffle equivalent to that provided by the stock pan. In some instances it will be necessary to mill away a portion of the short baffles. This will permit locating the pump pickup on the bottom of the pan to take advantage of the extra oil capacity. The cool oil lies closest to the radiating surface.

2. An undrilled boss for the addition of an oil-temperature sender.

3. Quieting—because the external fins and/or internal baffles make the pan a ribbed structure, it will not resonate as much as the unbraced stock item. Thus, your engine will be somewhat quieter.

4. Additional oil capacity up to three quarts more than stock. This is a mixed blessing on any engine other than one for all-out road racing purposes. Additional oil capacity, as already mentioned, does not lower oil temperature—it just increases the time required for the oil to attain a stable operating temperature, which is not usually of any real importance. Oil changes cost

Otto Parts finned oil pan and rocker covers are black anodized to improve cooling efficiency. Pan has 314 square inches of cooling surface area; covers add another 316 square inches.

Interior construction of Otto parts pan includes 24 posts which extend into hottest portion of oil to transmit heat to cooled surface. Baffle construction allows stock pickup location at coolest part of sump. Pan adds one pint to oil capacity.

more and the engine requires longer to cool off after it has been run. Additional oil capacity is helpful in a long race, providing the pickup is moved to the bottom to make all of the oil available for use. Additional capacity is insurance against losing an engine in the event that unexpected oil-consumption problems develop in a racing situation, or for engines on which the breathing problem has not been properly worked out. Big-capacity pans usually offer about 240 square inches of heat-radiating area, which is reduced to the equivalent of 24 square inches by surface polishing. In general, it can be said that an oversize pan is essential for a non-dry-sumped, road-racing engine.

5. Appearance. Most special aluminum parts are purchased for looks instead of function—that is, show, not go! Do not let your desires for a good-looking engine confuse you when final-assembly time arrives. Avoid polished fins!

Cragar oil pans provide additional oil capacity; may give ground-clearance problems. Baffles must be cut away to relocate the oil pickup on the bottom. Polished surface should be eliminated and pan black anodized for maximum cooling effect.

If you are tempted to retain them reread engineer Phil Irving's comments. If you already have such parts, "de-polish" them by sandblasting the polished surfaces. Black anodize the parts or cover the exterior surfaces with a thin coating of self-etching flat-black paint of the non-insulating variety. Black anodizing is a plating process that many plating shops can provide for you at reasonable cost. Incidentally, the black obtained by anodizing may turn out somewhat spotty as castings do not usually anodize perfectly. This will not impair the heat-removal characteristics. Although seemingly insignificant, these efforts will reward you with a cooler-running, longer-lasting engine. The same efforts should be applied to rocker-arm covers of the cast variety.

6. Better sealing surfaces. The machined surface of the pan that contacts the case bottom is much stiffer than the stamped steel original pan. The same is true for aluminum valve covers. Some are better than others. The stock valve cover gaskets have tabs that aid in retaining the gasket during installation. Some of the

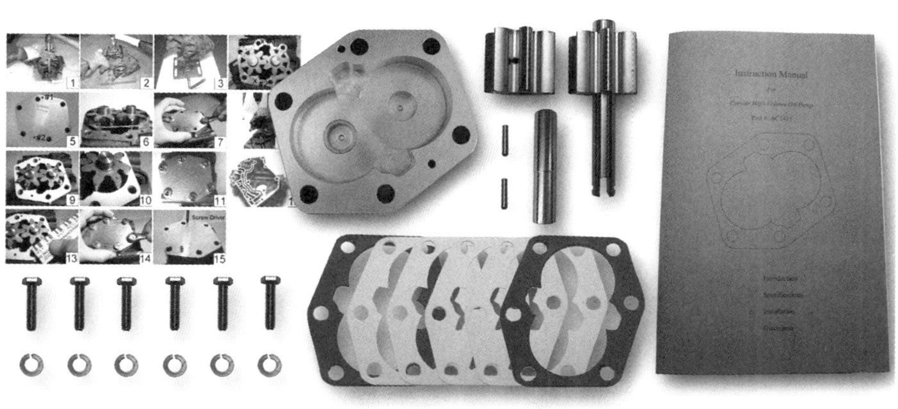

Shirok Enterprises of Colorado Springs, Colorado, makes this high volume oil pump kit. It is very useful if you are using a front-mounted cooler and filter.

Output volume of the Corvair oil pump is boosted one-third by addition of special Otto Parts pump cover to house Chevrolet 409 oil-pump gears. Installation can be made without removing engine from car.

aluminum valve covers have the spaces for the tabs.

Drawbacks? The pan will hang down beneath the already low engine. Ground clearance could be an issue.

Improving the Oil Pump

As has been mentioned previously, improving the flow of oil reduces bearing temperatures. This statement assumes that the oil-pump capacity can be increased to provide the additional flow. Fortunately, the Corvair's pump is located so that it lends itself to a simple modification for increasing its output volume. A spacer can be installed to house the ⅓ longer gears from a 409 Chevrolet thereby increasing output to 12 gallons per minute (GPM) at 4,000 RPM. This spacer is an aluminum casting that bolts directly to the stock accessory drive housing. Such an adapter casting, complete with gears, is manufactured by Otto Parts.

Although the 409 gears are 0.397-inch longer than the Corvair gears, the stock idler gear shaft can be used. The drive-gear shaft is relocated slightly to match with the Corvair's distributor-drive tang.

Oils

The Corvair needs the best oils to provide good viscosity under the extreme conditions of temperature encountered in an air-cooled engine, especially a modified one. Multi-grade oil acts like a 20-weight to reduce cold friction and retains the viscosity of a 40-weight oil as the temperature is increased. Shell "Rotella" oils, in various grades, have worked well in the Corvair. Don Eichstadt and other Corvair racers report good experience with D-A Speed Sport oils. There are several synthetic oils that work great in the Corvair, RedLine, Amsoil, and Mobil1 all have their proponents. The high-temperature capabilities of the synthetics are perfect to help the Corvair cylinder head temps in competition usage. If you decide to go with a synthetic oil in your new motor, I suggest you run the motor on normal non-synthetic oil for the first 500 miles or so. The extreme lube capability of the synthetic may prevent certain new parts, like piston rings, from mating with their counterparts, resulting in long-term oil consumption. There are still some non-synthetic oils that perform very well. Penn Grade Oil, from Pennsylvania sources is competition proven, it was previously marketed as Kendall Oil, a favorite of racers, but comes from the same wells and factory as the original Kendall. No matter which oil you choose, always change your oil when it is hot to get the best flushing action. Avoid oil additives as they may not be compatible with the additives that the oil manufacturer has used.

Internal Oil-Sump Baffling

Reference to the need for internal baffles in the sump has already been made in the discussion of special pans. Baffles extending to the underside of the crankcase separator wall, which separates the rotating parts from the oil supply, assist in drawing heat from the oil to the cooled cover and reduce oil from sloshing away from the pickup. In addition, still another set of baffles is needed for hard cornering as is experienced in autocrossing or road racing. While the pushrod tubes are protected from receiving oil from the crankshaft,

Splash baffles prevent loading rocker covers with oil in cornering with possible starvation of pump. Holding screw goes between lifter bores, exits inside crankcase. Use 1-½ inch 8-32 screw into tapped hole.

Two crankcase cover oil separators. Cover at left was introduced on Spyders in '62, was soon made standard throughout engine line. It reduces oil pull over into the positive-ventilation breathing system. Additional opening area is the secret of its success.

rods, and etc., there is nothing to protect them from receiving a literal flood of oil when cornering sloshes oil up the side of the crankcase walls from the sump. When oil rushes into these tubes to fill the rocker covers, temporary starvation of the pump can occur because of the quantity of oil that has left the sump. And, the valve-stem oil seals are swamped with more oil than they can handle. Elimination or reduction of the problem is easily accomplished by attaching short strips of aluminum to the separator wall to protect the tube openings from sloshing oil. One or two rivets will suffice for attachment of each shield. Leave 1⁄32 to 1⁄16 clearance at the outer edge next to the crankcase wall to permit normal drainage of the tubes. Make six of the shields from 0.030 or heavier aluminum, 1-3⁄16 × 3-1⁄8. Don Eichstadt recommends the use of a single six-inch-long vertical baffle on either side of the pan to reduce the need for the splash baffles.

Doug Roe added an extra piece of sheet metal to the bottom of his oil pickup so that the oil about to enter the pickup will not be forced away from the opening during cornering. For the last 20 years, I have been running the Otto Parts pan, the Otto Parts baffle spacer and the companion pickup that drops the oil entry down to the very bottom of the sump. The baffle spacer prevents the oil from sloshing up into the tubes.

Doug Roe's oil pickup modification.

Crankcase Breathing

Corvairs made since 1963 include positive crankcase ventilation equipment. Models prior to that year could be so equipped, especially in certain states (California) that were working mightily to correct the smog problem. These systems effectively remove harmful, oil-contaminating vapors from the crankcase, thereby preventing corrosion and sludge formation. The use of these systems is very desirable on engines that are used for short trips or stop-and-go city driving. Keep the system clean and it will not, contrary to common fictions, reduce performance at all. In fact, taking it off can foul up the idle mixture if the system is coupled through a valve to the intake manifold.

Engines running in competition need additional crankcase breathing capacity to avoid the development of internal pressures which will blow oil past the seals and gaskets which close the engine. Of three specific approaches that have been used, two definitely do not work and should be avoided. If you want to cover the side of your engine with oil, mount one or more "big" breathers on each rocker-arm cover. Regardless of internal baffling or special plastic stuffing, no breather can hold back the flying oil mist which is inside of these covers. The valve covers contribute to the cooling of the engine because they are constantly bathed in a thin film of oil and the outer surface is exposed to cooling air. Adding a breather to the cover reduces the area available for cooling. The second non-working approach is the addition of a baffled tube to the oil-filler pipe. This tube is the recipient of high-velocity oil spray slung off the crankshaft. If you do not believe it, just take off the cap while the engine is running. Wear your old clothes and have a can of Gunk handy to clean up the engine and its compartment! The stock breathing tube cannot be used by itself, of course, because its tiny 3⁄8-inch I.D. creates high velocities of the escaping air/oil mist to carry out oil quite handily. The same is true of the dipstick tube that must be carefully gasketed or even plugged for competition to prevent oil from spewing out of that opening.

The reason all of the foregoing systems cannot work at high RPM is the fact that they provide openings with very small area. A satisfactory installation must provide adequate area to slow down the mist and sufficient length to permit the condensed oil to collect and drain back into the engine. Drag racers have the answer with their catch-tank breather. A one- or two-gallon catch tank has a baffle in the center, with outlets in the form of two to three oil-filler-type breathers. Two 1-inch or larger rubber hoses connect the tank to the engine, often to the front of the rocker-arm covers. A drain is provided in the tank so the oil can be removed at the end of each run. I ran a pair of large tubes, one from each side of the upper baffle, along with the stock tube, into a large breather up under the rear window. The breather drains back into the top of the valve covers. To meet the racing requirements, the system ends in a catch tank.

13

Clutch and Flywheel

Corvair Clutch Specifications

Not all of the following clutches are interchangeable. The 1964-later models with bent-finger design must be used with a stepped flywheel and other '64 parts when installed into earlier models.

As you can see from the table below, changes occurred over the 1960–64 period. The latest Spyder or truck clutches (1964 and later) are obviously best, with their improved torque rating and high-speed shiftability. In comparison with the first 1960 clutches, '64s have ⅓ more spring pressure.

The early tendency to resist releasing at high RPM has probably lost more races than Corvair owners care to relate. In addition, large hunks of friction material have often been lost out of the earlier clutch discs. Chevy's first step to reduce the tendency to "hang up" was to add a clutch-pedal assist spring. In '61 the disc diameter was reduced to 8 inches. This may have improved the shifting slightly, but it reduced the torque ratings of the clutch. Changes in 1962–63 eliminated the clutch-assist spring, and it must always be removed when installing a later-model clutch into a '60–'61 model. Real improvement in high-speed shifting was accomplished in 1964 by changing to the bent-finger diaphragm. According to Chevrolet, the bent finger design positions the mass of the release fingers to ensure a positive engaging force for high-speed shifting. This is the same type of clutch used on the Chevy 409, in a smaller size. The '64 and later style clutch is best for modified Corvairs, preferably the Spyder/140/180 because it has better pressure-plate material. All of the components should be matched: clutch cover, flywheel, disc, fork assembly, pivot ball, and the bearing.

The new parts can be installed in 1960–63 flywheel housings, when the '64 crank is used, providing the seal is moved 0.070-inch rearward to mate with the revised sealing faces on the hub. If the '64 clutch assembly is used with a '60–'63 crankshaft you might have to do some minor relieving of

Year	*Rated Torque Capacity (lbs. ft.)*	*Diameter (in.)*	*Area (sq. in.)*	*Pressure (lbs.)*
1960: 80 & 95 HP	151	9.12	71.8	900–1,050
1961: 80 & 98 HP	138	8.0*	44	900–1,050
1962: 80 & 102 HP	141	8.0*	44	900–1,075
1962: 150 HP	168	8.0*	44	1,050–1,250
1963	Same specifications as 1962: 80, 102 & 150 HP			
1964–69: 95, 110 HP	197	9.12	71.8	1,250–1,450
1964–69: 140, 150 & 180	218	9.12	71.8	1,275–1,475

*All trucks and forward-control vehicles used 9-inch clutches in every year. Nominal pressures for 1963 trucks are 1,000–1,200 pounds; 1964 trucks are 1,400 pounds. Pressure plates of the 140/150/180 HP models are made of nodular malleable iron, others of cast iron. All pressure plates can be used with either the 8- or 9-inch discs. Flywheel weight is approximately 19 pounds.

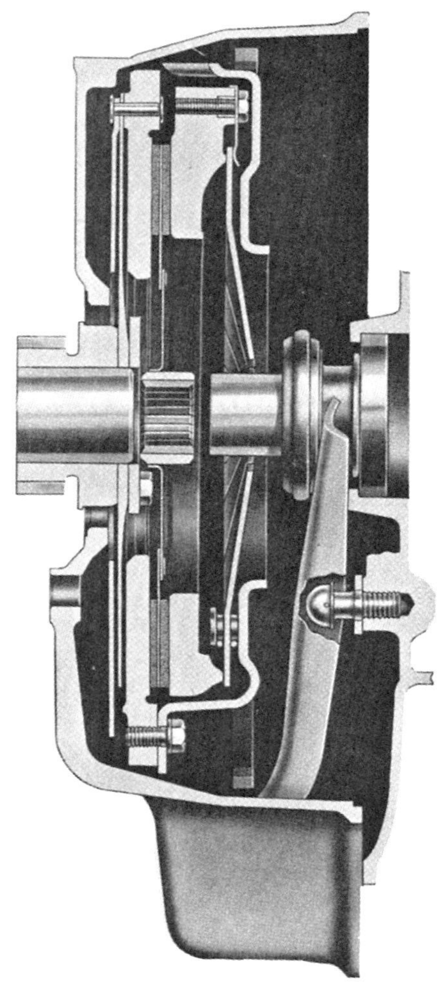

Bent-finger-design clutch introduced in 1964 reduced high-speed-shifting problems. Unit requires stepped flywheel for installation into 1960–63 cars.

the housing by grinding. Any new assembly, even factory parts, should be dynamically balanced prior to use.

In 1965, the 140 HP motors and 180 Turbos used a special HD pressure plate. They can be visually identified by the extra material hanging out from the contact area inside the clutch cover. They were also built of a better "Nodular iron" material. The reasoning behind the HD clutch is the decreased low-end torque of the 140/180. With the decreased torque, those vehicles would need extra inertia to accelerate from a stop. Once spun up from idle, the heavier clutch/flywheel assembly would get the cars away from the stop-light without bogging. For any racing-type application, except drag racing, the HD pressure plate is not needed, although the better material might make it a nice "donor" piece for weight reduction techniques.

Another problem with the Corvair has been starter ring gear warpage. The cause for this is flywheel run out due to loose rivet problems, note following. One symptom of this problem is broken noses on the starter. Because the ring gear is skip-welded to the clutch cover, and the clutch is bolted to the flywheel, when the flywheel rivets loosen up, the clutch cover run out increases and the ring gear with it. If the engine stops and the ring gear is at maximum outward travel, the next engagement of the starter can do damage to both the starter (resulting in a broken starter nose) and the ring gear, causing bent and warped teeth. Before investing too much in refacing and rebuilding a clutch, check the condition of the starter ring gear. Look for broken welds and bending of the starter ring. The starter ring is replaceable but it may cost more than the rebuilt clutch.

Discussions with several specialty clutch houses indicate that their best answer to Corvair clutch problems is to use the truck pressure plates or to switch to the '64 parts. There are several special Corvair clutches available. A high-pressure diaphragm spring can be inserted in the original cover. Aluminum pressure plates were available for the Corvair, complete with metal-sprayed coatings on the contact faces, but none have been produced in years.

The original asbestos lining used on the Corvair clutch discs is no longer available. Unless you run into an original disk from storage, you will be buying a different lining material. Unlike some of the replacement brake shoes, however, several replacement clutch materials have proven successful. There are Kevlar material facings available, as well as semi-metallics. Do not run the full metallic discs on street cars. Aside from being a pain to drive, they will wear out prematurely, because slippage is not their friend.

Several people have used the spring-center Vega-style clutch disc in the Corvair. Although close to the Corvair in many dimensions, the spring center portions can contact bolt heads on the crank if not fully clearanced during assembly. There are also replacement disks that use a separate center with large rivets connecting it the outer disk. They have also had clearance problems. The welded center clutches are the safest choice. They clear all parts and are closest to the original design.

Stock Flywheels

Corvair flywheels are unusual in that they are of three-piece riveted construction. That portion which bolts to the crank hub is a tapered steel spring that absorbs bending resonance. As the photos show, the outer fingers of this spring are riveted between a steel ring and the cast-iron facing. This flywheel will run true, even when slightly misaligned. The flywheels often become loose and rattle. This is because the rivets sometimes work loose. They could have been reriveted, but the dealers always replaced them. You can usually rework a flywheel, installing bolts in place of the rivets, prior to having it balanced, as there is seldom anything wrong with the working surface. There are several suppliers who specialize in bolting these flywheels. The advantage of seeking them out, instead of doing it yourself, is the fixtures they use, not only for alignment during the assembly, but for checking lateral and radial run out when done. One of the things which may aggravate this loosening of the rivets is the use of a loose fit between the flywheel bolts and the flywheel mounting holes. Bolts are $\frac{11}{32}$ fine thread—a rare thread, but the mounting holes in the flywheel are almost $\frac{3}{8}$-inch diameter. Either build up the bolts under the heads so that they will be approximately $\frac{3}{8}$ inch, or use $\frac{3}{8}$-inch aircraft bolts that have been threaded to within 0.140 inch of the head plus the thickness of any wavy lock washer being used. This later course will require drilling and tapping the crank hub for $\frac{3}{8}$ inch-24 threads. Tools are available to do this at home. The net effect is to dowel the flywheel and prevent any movement under hard acceleration and deceleration. Stock bolts (safety wiring optional) are usually adequate. On the 1964–up flywheels, if you do need to resurface the flywheel contact surface, be sure to maintain the difference in height between the contact surface and the bolt-down surface for the pressure plate. Several other Chevrolet products of the 1960s also used a step-faced flywheel, so the machine shop should know what to do.

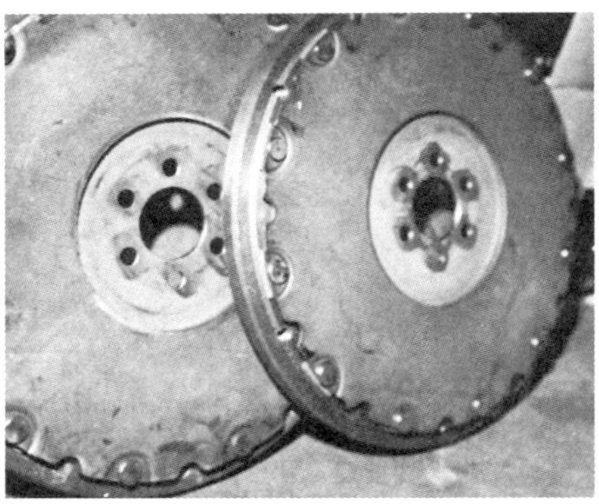

Rear of Corvair flywheels are similar though '64 and later step-faced flywheel at right is immediately obvious from clutch side. Three-piece-construction flywheels have flexible center-plate mounting.

Special Flywheels

Strangely enough, the 19-pound Corvair flywheel may be too light, especially for drag-race competition where it is essential to break the rear wheels loose for best off-the-line acceleration with the available rear-end ratios. When the tiny Corvair engine is forced to pull 2,400 pounds off the line, it needs more kinetic energy than the stock flywheel can store. More kinetic energy can only be obtained by (1) higher initial RPM with severe clutch slipping, or (2) a heavier flywheel with more-moderate revving of the engine as the clutch is engaged. A heavy flywheel stores kinetic energy. Use of heavier-than-stock flywheels for drag racing has been going on for several years. So, if you install one, you are not pioneering new territory. And, while the heavier flywheel may make engine acceleration a tiny bit slower, you get back as much or more than has been lost when you shift. The engine won't "bog down." This is really important in drag racing. Some Corvair racers are experimenting with steel flywheels weighing more than 35 pounds—with gratifying results.

That the Corvair has a special need for added kinetic energy can be seen by the many attempts to use smaller-diameter tires to overcome the too high gears which must be used in the differential (3.89:1 is the lowest you could get, stock). Several cars have been seen with 12-inch midget wheels on the rear. This is a good answer to the problem, providing the association or strip rules permit 12-inch wheels. Some will not allow wheels smaller than 13-inch diameter, good luck with the tires. A heavy flywheel is helpful with either wheel size, but even more with the 13-inchers.

Light-weight flywheels can be helpful to obtain quicker acceleration out of corners for road racing or autocrossing (slaloms). One Corvair racer "scallops" away the inertia ring and steel backing plate between those pairs of rivets which do not straddle a clutch mounting hole. The inertia ring can have its OD turned slightly, also. This same racer claims that the standard clutch (rather than the high-performance version) should be used because the springs are the same but the standard version is lighter. You will want to check all critical dimensions before assuming that the same spring will provide the same pressure on the disc. Aluminum flywheels are often advertised with imaginative but technically incorrect claims. The best reason for using an aluminum flywheel is that cast-iron flywheels have been known to disintegrate, even when banded with a steel ring gear—which the Corvair is not. I have never seen a Corvair flywheel disintegrate, I have experienced it on a Chevy V8.

Ken Hand is developing a multi-plate clutch assembly for the Corvair. The clutch-mounted starter ring gear has always been an impediment to the adaptation of other clutches into the Corvair powertrain. Several one-off conversions have been done by racers. Depending on the cost and material selection, this may turn out to be the best racing option and, perhaps, a high-horsepower street option as well.

14

Transmissions, Gearing, and Axle Ratios

The Powerglide has a storied history in drag racing, both in Corvairs and in tons of other cars. Two items you should address for drag racing the PG. First, you will have to look around a bit, but companies can provide torque converters modified internally for high stall speeds. This helps the Corvair leap off the line. Second, that loose converter does tend to heat the trans fluid up a bit. All forward control Corvairs used an external cooler on their Powerglides. Those can be adapted to the car PG. The cooler is fed off the front of the trans and a couple of small modifications to the front crossmember must be made to connect it, as well as an internal modification to force the fluid through the cooler.

Unless you are dollar limited to using a 3-speed transmission that was originally installed in your car, you will be running some variation of a 4-speed transmission in your Corvair race car. There were three basic 4-speed transmissions used in Corvair cars. All were built at a GM plant in Saginaw, Michigan. The early style, used from 1961 through 1963 is lighter duty and shared a few parts with the Pontiac Tempest of the era. Some internal parts were upgraded for the 1964–65 version. Even though the body style changed considerably from 1964 to 1965, the 4-speed transmissions are interchangeable between the two. In 1966, everything changed. A new transmission based on the stronger big car 4-speed replaced the older styles. The reason that the 1966 through 1969 manual transmissions have been referred to a "Saginaw" transmission was their design similarity to and some parts interchangeability with the conventional application (non-Corvair) "Saginaw" transmissions. Those were called Saginaws to differentiate them from other 4-speed transmissions, such as Muncie and Borg Warner. The '66–'69 style is based on a 4-speed design used in many other GM cars then and since. Many internal parts are shared with those transmissions. The big benefit of this is that replacement gear sets from other cars can be installed for the correct racing application. Yenko Chevrolet supplied the close-ratio transmission for the Yenko Stingers and many others have been building them up since then. These are not for street use, because the 2.56 to 1 low gear is much harder on the clutch and the engine than the original 3.11 to 1 stocker. But with the right gearing, the close-ratio box can provide the autocrosser or track racer with one (or even two) usable additional gears. Spend the time to calculate speeds in gear with the transmission gears available, combined with the commonly available axle ratios, 3.55 to 3.27, and tires that meet your needs. Time spent here pays dividends on the track. A couple of preparation notes are important. The 4 or 6 bolts that hold the transmission and differential together can loosen in use. Safety wiring of the bolt heads is recommended. If you have the transmission out of the car, the seal around the shifter shaft should be checked for leaks.

Chevy offered a limited-slip differential for the Corvair, the early design fits through 1964, the late style through '69. The "Positraction" axle has a completely different ring gear carrier assembly inside. To convert a differential to Positraction you need the special yoke and the special adjustment sleeve for the passenger's side, as well as the special posi carrier. If you have the transaxle out of the car, be sure to check the throwout bearing sleeve/collar. Sometime, during the life of the car, the engine and transaxle may be been separated carelessly. If the transaxle is not removed carefully, the input shaft will bind in the end of the crank, or the clutch disc, and the shaft will pull out from the trans as the two are separated. If one or the other of these drop down, the shaft is pulled sideways in the sleeve/collar and the collar unit cracks. There are two bad effects from this, the throwout bearing loses full travel, because is sticks at the split area, and the collar will begin to leak lubricant through the crack in the collar. This will kill your clutch disc, and things will begin to slip. The good news is that a new, replacement collar assembly is available that has been redesigned to accept a pair of the clutch seals, one on top of the other. The bad news is that the collar is the last thing to come out of the differential housing during disassembly, so a complete disassembly is needed to install it.

Corvairs are over-geared for drag-race acceleration. The best ratio for acceleration has not been supplied in Corvair passenger cars since 1962—the 3.89 to 1 gear set. This is the lowest ratio that was ever factory supplied for the Corvair. It can be installed in the '63–'64 differentials without modification. It will also fit into the '65–'69 differentials by merely grinding away any interfering ribs, then drilling and spot-facing three additional holes so that you end up with six holes that mate with those in the ring gear. While you have the transaxle out to install the lower gears, you may want to consider updating your trans, several factory revisions were made to the Corvair transmissions to improve shifting ease, etc. Because the 3.89 ratio is too high for the tiny Corvair engine to give its best acceleration performance, other tricks must be used to get the ratio still lower. For early models, one simple way is to install short tires on Chevy II wheels Part No. 3790690. Not only are these tires approximately one-inch smaller in diameter than the stock Corvair, the narrow-base wheels (4.0-inch rims) help to reduce traction so the wheels can be "broken loose" at the start. This was the setup used by Dick Griffin in setting his 1962 NHRA drag record. Tires pumped to 50 or 60 psi improve the situation by reducing the traction so the engine can get wound up more quickly to its best RPM. Low-profile 13-inch tires on wide rims will lower the overall ratio to an effective 4.11:1. And, unlike other tires that you might buy for dragging, these can be safely used on the street for normal driving. The total outside diameter of these will be about 21 inches installed. Several runs of aftermarket 4.11 gears have been available over the years. There have been hardness problems with some of them. All the aftermarket gears have been drilled for both the 6-bolt and 9-bolt patterns, so fitting a particular axle is no longer a problem.

One mistake that is easily made by the beginning drag racer is the tendency to gear so that the engine will be turning that RPM at which maximum HP is produced at the end of the strip. If your tests on the dyno show that the engine produces its maximum at 4,800 RPM, select a gear that will allow the engine to turn past its peak HP speed, even though the engine will be somewhat "down" the HP curve at the end of the course. Better elapsed times will result. If you are checking the car for acceleration, work for the best E.T. over the quarter mile, regardless of the fact that this may produce a slower end-of-trap speed. It is the gear that gets to the other end first which is awarded the trophy,

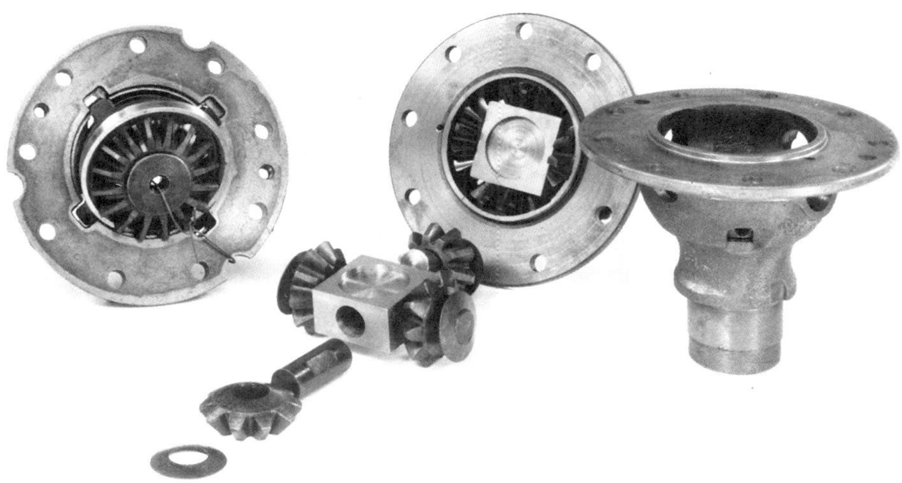

Essential modification for any V8 Corvair is the addition of two more spider gears to the stock Positraction unit. Clark's sells the parts for the conversion.

AXLE RATIO TO ROAD SPEED

Gear	*Transmission, Year, and Ratio*	*Axle Ratios*	*Overall Ratios*	*Theoretical MPH @ 1000 RPM*
1st	4-speed (1964–65)	3.27:1	10.46:1	6.7
	3.20:1	3.55:1	11.36:1	6.2
	4-speed (1961–1963)	3.27:1	11.93:1	5.9
	3.65:1	3.55:1	12.96:1	5.4
		3.89:1	14.20:1	4.9
2nd	4-speed (1964–65)	3.27:1	7.13:1	9.8
	2.18:1	3.55:1	7.74:1	9.1
	4-speed (1961–1963)	3.27:1	7.68:1	9.2
	2.35:1	3.55:1	8.34:1	8.4
		3.89:1	9.14:1	7.7
3rd	4-speed	3.27:1	4.71:1	14.9
	1.44:1	3.55:1	5.11:1	13.8
	(all years)	3.89:1	5.60:1	12.6
1st	3-speed (1960, 1964–65)	3.27:1	10.53:1	6.7
	3.22:1	3.55:1	11.43:1	6.2
		3.89:1	12.53:1	5.6
	3-speed (1961–1963)	3.27:1	11.45:1	6.1
	3.50:1	3.55:1	12.43:1	5.7
2nd	3-speed (1960, 1964–65)	3.27:1	6.02:1	11.7
	1.84:1	3.55:1	6.53:1	10.8
		3.89:1	7.16:1	9.8
	3-speed (1961–1963)	3.27:1	6.50:1	10.8
	1.99:1	3.55:1	7.06:1	9.9
		3.89:1	7.74:1	9.1
High	All years	3.27:1	3.27:1	21.5
	1.00:1	3.55:1	3.55:1	19.8
		3.89:1	3.89:1	18.1

Chart above, from *Corvair Communiques*, is for passenger cars with 6.50-13 tires rated at 853 revolutions per mile. Tire differences can make chart figures inaccurate by up to 6%. 1965 and later gear ratios and transmission ratios are the same as the '64s.

'66–'67 three-speed ratios are 3.11, 1.84 and 1.00, four speed has 3.11, 2.20, 1.47 and 1.00. Both are all-synchro with same-size gears as used in big Chevys.

even though the winner may be running slower speeds at the end than the loser.

For autocross, the rear-axle ratio selected should take into account the courses usually encountered. The 3.27 ratio with the '64–'65 four-speed transmission ratios will give the least "gap" at the shift points, helping to keep the engine from "bogging down" as you shift. This makes a three-speed box almost useless for autocross or drag racing. This 3.27 ratio might be very desirable on a short, twisty course, but better performance on a high-speed course might benefit from the use of the 3.55 or 3.89 gear to provide better "pull" in 4th gear.

Top-speed gear selection is a different matter. The highest top speed is not always obtained with the 3.27, Corvair's highest gear ratio. Higher top speeds are often obtainable by letting the engine "twist tighter" with a 3.89 or 3.55 gear—usually the latter. If a perfectly tuned engine will not turn somewhat higher RPM than those where maximum HP is produced, then a lower gear ratio will produce higher top speed. Again, Chevy II wheels can be called into service. This time you can get the 14-inchers. If your 3.89- or 3.55-equipped axle will pull larger diameter tires without any bogging down, then you can calculate whether you will be better off to use the larger wheels and tires temporarily instead of changing the rear-axle ratio. Top speed is so seldom needed for everyday driving that the average Corvair owner will be able to obtain the best all-around performance by installing the 3.89 set in the rear end and concentrating on through-the-gears performance. It does tend to drone on the highway, however. You might consider a second pair of rear wheels and tires—one pair to drive on, one pair to drag race on.

Testing—get your engine running with close to the right jet and ignition settings and install the gear ratio that seems to fit your requirements best. Find a long smooth stretch of lightly traveled road. Take along a couple of five-gallon fuel cans to use as markers, placing these about ¼ mile apart. A notebook, pencil, and stopwatch will also be required. Record every step of your testing or you will forget the figures. Make runs through the gears, stop-watching the 0–60 MPH times and changing shift points. (Safety first!) When you find the low-to-second shift point that gives best acceleration, work on the second-to-third shift point. The high-gear shift point will probably occur at the RPM equivalent where you found the best third-gear shift point occurring. With three-speed transmissions shift points are more important because of the wide gaps between the ratios. Once you have established the best shift points, record the acceleration times which you obtain from 0–50 MPH, 0–60 MPH, and 30–80 MPH. Keep these records for reference so that you will be able to make future checks of engine performance. These records will show you whether additional changes are helping or hindering. Now that our smart phones contain GPS tracking, and electronic devices are so inexpensive, you can choose any number of ways to track your progress.

15

Tuning

Most modified engines never produce the maximum HP outputs of which they are truly capable. Why? Because their builders seldom spend the final efforts required to make their creations run perfectly. Make your investment pay maximum dividends through careful attention to details during assembly, then tune it to perfection after installation. The $200 to $250 worth of chassis dyno time that you purchase will be worth far more in terms of the performance that you will obtain as a result. Items that deserve special attention in tuning include carburetor jets, linkage and synchronization, ignition timing (both initial setting and centrifugal-advance rate and amount), sparkplug selection, and choice of transmission and rear axle gear ratios.

Surprisingly, multiple-carburetor installations do not stay synchronized and an occasional check with a Uni-Syn gauge, or similar tool, is essential. Prior to tuning or competition, check to make sure that the carburetor linkage works easily and closes the carbs positively when pressure is removed from the foot pedal. You must check to see that the throttles open fully and that gaskets are not obstructing the throttle bore. The 1965 Shop Manual has a great section on setting up the 140 throttle linkage, but the Uni-Syn or the newer design Synchrometer can be used to easily balance the idle airflow. Correct jets can be selected by running the engine at wide-open throttle on a chassis dyno with an exhaust-gas analyzer, or on the road with a wide-band oxygen sensor air/fuel gauge. If it can be carried in the car, it can also be used to check accelerator-pump settings, even though many of these instruments exhibit a several second response time lag.

These units still work fine for airflow measurement. If you have tall vent tubes on your carburetors, a tubular adapter could be used to seal the top of the carb to the Unisyn.

When you have installed the jets required for maximum power (12:1 air/fuel) you may discover that this setting gives less economy than you prefer for ordinary driving. If this is the case, use leaner jets for around-town and highway driving, remembering that the engine is being fed a too-lean mixture for full-throttle conditions. Rejet the carburetors when you want to use full power. Part-throttle or "cruise" can exist on 14:1 ratios. The wide-band air-fuel sensor is the quickest path to selecting the proper jets, but plug color is also a good indicator of mixture conditions. Plugs however, especially when new, take time to color. Plug-color readings are only meaningful when the engine is declutched and the ignition cut at the end of a high-speed, full-throttle, high-gear run. If the car is allowed to slow down with the engine still running, plug readings become meaningless. Plug readings can be made after full-throttle runs on a dyno with the car being operated in an intermediate gear, but the same does not hold true for road tests.

This is the newest airflow management tool. It can be used to determine the "relative" airflow for multiple carburetor systems. With these adapters, the Synchrometer is used to verify idling on both primary carbs. The idle speed screws can be adjusted to equalize the airflow.

Air cleaners can affect mixture, so run the dyno or road test with the air cleaners installed. Dyno tests can show whether there is any restriction due to the air cleaners.

Plug Readings

Rich	Sooty or wet plug bases, dark exhaust valves.
Correct	Light-brown color on porcelains, exhaust valves red-brown clay color. Base of plug slightly sooty (should leave slight soot mark on hand when plug base is turned against palm).
Lean	Plug base ash gray. Glazed-brown appearance of porcelain may also indicate too-hot plugs. Exhaust valves whitish color.

Corvair Engine Specifications and Tune-Up Data

1962

TURBO-AIR ENGINES 6 Cyl. Horiz. Opposed 145 Cu. In. Displacement	80 HP	84 HP (Monza w/Powerglide) / 102 HP (RPO 649)
Gross Torque (Ft.-Lbs.)	128 @ 2300 rpm	(Monza)— 130 @ 2300 rpm (RPO 649)— 134 @ 2900 rpm
Compression Ratio	8.0:1	9.0:1
Rated GHP Developed at	4400 rpm	
Bore and Stroke	3.44" x 2.60"	
Firing Order	1-4-5-2-3-6	
Test Compression	130 psi.—(20 psi. max. variation)	
Spark Plugs	AC 46 FF	AC 44 FF
Gap and Torque	.035"—20-25 ft. lbs.	
Distributor Dwell	31°-36°	
Point Gap	.019" (New) .016" (Used)	
Point Tension	19-23 oz	
Ignition Timing (BTDC)	4°-10° (Synchro.) 13°-18° (Auto. Trans.)	13°-18°
Lifter Adjustment	1 Turn Down from "No Lash"	
Fuel Pressure	4-5 psi @ Idle to 1000 rpm	
Engine Idle RPM	500	600 w/Synchro. 500 Auto. (In Drive)
Crankcase Capacity	4 Qts. (4½ Qts. w/Filter Change)	

1963

TURBO-AIR ENGINES 6 Cyl. Horiz. Opposed 145 Cu. In. Displacement	80 HP	84 HP (Monza w/Powerglide) / 102 HP (RPO L62)	150 HP Turbocharged Monza Spyder (RPO L87)
Gross Torque (Ft.-Lbs.)	128 @ 2300 rpm	(Monza)— 130 @ 2300 rpm (RPO L62)— 134 @ 2900 rpm	210 @ 3300
Compression Ratio	8.0:1	9.0:1	8.0:1
Rated GHP Developed at	4400 rpm		
Bore and Stroke	3.44" x 2.60"		
Firing Order	1-4-5-2-3-6		
Test Compression	130 psi.—(20 psi. max. variation)		
Spark Plugs (Prod.)	AC 46 FF	AC 44 FF	
Gap and Torque	.035"—20-25 ft. lbs.		
Distributor Dwell	31°-34° (33° preferred)		
Point Gap	.019" (New) .016" (Used)		
Point Tension	19-23 oz.		
Engine Normal Idle	500 rpm	600 rpm— w/Synchro. 500 rpm— Auto. (In Drive)	850 rpm
Ignition Timing (BTDC Settings at Normal Idle)	*4°-10° (Synchro.) *13°-18° (Auto. Trans.)	*Range 13°-18° (Prod. set at 13°)	24° (Do not Exceed)
Lifter Adjustment	1 Turn Down from "No Lash"		
Fuel Pressure	4-5 psi @ Idle to 1000 rpm		
Crankcase Capacity	4 Qts. w/o Filter Change—4½ Qts. w/Filter Change		

*Disconnect vacuum advance line and cap opening at manifold.

Set the ignition timing per factory specifications, using added initial advance with caution as discussed in the "Ignition" chapter. Avoid setting the ignition timing while the engine is being tested on a chassis dynamometer—unless you make readings that are held long enough for temperatures to stabilize. Flash readings made in conjunction with quick twists of the distributor to get impressive readings are not the answer. If you do this and then run the engine with the same setting, especially at high speeds on the road or in competition, you can destroy the engine. I happen to know this from bitter personal experience. What happens is this: the spark setting that can be tolerated by the engine during a flash reading before cylinder head and sparkplug temperatures have stabilized is more advanced than the engine can use after temperatures have stabilized at their higher values. As a result, the dyno setting for a flash reading will always be too far advanced! Using this setting causes destructive detonation.

Tuning pays off in performance dividends. This is true with any automobile, including stock or highly modified Corvair powerplants. One careful tuner, Don McKenna of San Jose, California, made a habit of tuning his Corvair on the way to and from work. He selected a length of road without traffic and used his stop watch to check out the car's performance. Temperature effects are equalized by only testing on cool days, of which there are many in northern California. Don has made available the test results from two

1960 CORVAIR COUPE TESTS—BY DON MCKENNA

Run	*Condition*	*0–60 MPH*
1.	Stock, 0.050 jets, with muffler and air cleaner	14.6 seconds
2.	Above, plus AC44FF plugs set at 0.030, points at 0.014, no air cleaner, 0.055 jets, 1961 high-performance muffler	13.1 seconds
3.	Above, plus 1.2:1 fan pulley	12.5 seconds
4.	Above, plus headers with two stock mufflers	11.6 seconds
5.	Above, plus 0.060 jets	11.2 seconds

Corvair Engine Specifications and Tune-Up Data (cont.)

1964

TURBO-AIR ENGINES 6 Cyl. Horiz. Opposed 164 Cu. In. Displacement	95 HP	110 HP (RPO L62)	150 HP Turbocharged Monza Spyder (RPO L87)
Rated GHP Developed at	3600 rpm	4400 rpm	4000 rpm
Gross Torque (Ft.-Lbs.)	154 @ 2400 rpm	160 @ 2600 rpm	232 @ 3200
Compression Ratio	8.25:1	9.25:1	8.25:1
Bore and Stroke	3.44″ x 2.94″		
Firing Order	1-4-5-2-3-6		
Test Compression	130 psi.—(20 psi. max. variation)		
Spark Plugs (Prod.)	AC 46 FF	AC 44 FF	
Gap Set	.035″	.030″	
Installation Torque	20-25 ft. lbs.		
Distributor Dwell	31°-34° (33° preferred)		
Point Gap	.019″ (New) .016″ (Used)		
Point Tension	19-23 oz.		
Engine Normal idle	500 rpm	600 rpm—w/Synchro. 500 rpm—Auto. (In Drive)	850 rpm
Ignition Timing (BTDC Settings at Normal Idle)	*2° (Synchro.) *10° (Auto. Trans.)	*12°	24° (Do not Exceed)
Blower Belt Strand Tension	**New Belt 70-80 lbs. Used Belt 50-60 lbs. (Set with Strand Tension gauge)		
Hydraulic Lifters	Prod. Set 1 Turn Down from "No Lash"		
Fuel Pressure	4-5 psi @ Idle to 1000 rpm		
Crankcase Capacity	4 Qts. w/o Filter Change—4½ Qts. w/Filter Change		

* Disconnect vacuum advance line and cap opening at carburetor.
** Periodic retensioning is not recommended. Retension only if belt slippage occurs.

1965

TURBO-AIR ENGINES 6 Cyl Horiz. Opposed 164 Cu. In. Displacement		95 HP	110 HP	140 HP	180 HP Turbo-charged
Rated GHP developed at		3600	4400	5200	4000
Gross Torque (Ft.-Lbs.)		154 @ 2400	160 @ 2800	160 @ 3600	265 @ 3200
Compression Ratio		8.25:1	9:1	9:1	8:1
Bore and Stroke		3.4375" × 2.9375"			
Firing Order		1-4-5-2-3-6			
Test compression		130 PSI			
Spark Plugs	Standard	AC-46FF	AC-44FF		
	Colder	AC-44FF		AC-42FF Competition	
	Gap Set	.035"	.030"		
Installation Torque		15–20 lb. ft.			
Distributor	Point Dwell	31°–34°			
	Point Gap	.019 (New) .016 (Used)			
	Arm Spring Tension	19–23 ounces			
Condenser		.18–.23 MFD			
Engine Normal Idle	Synchromesh	450-500	600-650		850
	Automatic	As low as possible for smooth idle			
Ignition Timing (BTDC)*	Synchromesh	4°–8°	12°–16°	16°–20°	24°
	Automatic	12°–16°	12°–16°		
Blower Belt Strand Tension		55 +/– 5 Lbs. (Used) 75 +/– 5 Lbs. (New) Using Strand Tension Gauge			
Hydraulic Lifters		Prod. Set 1 Turn Down from "No Lash"			
Fuel Pressure		4–5 Lbs. Idle—1000 RPM			
Crankcase Capacity		4 Quarts + ½ Quart filter = 4-½Q total (with stock pan!)			

* Disconnect vacuum advance line and cap opening at carburetor.

Corvairs he has owned, a 1960 with 3-speed, and his 4-speed '63 Spyder. Both of these cars have been very successful in autocross competition and were equipped with the 3.55:1 rear axles. As you can see, these methodically obtained records proved that some changes "work" and others do not. It is often tempting to tune by the seat of your pants, but stopwatches and dynamometers are better producers of results that bring home the trophies.

1963 CORVAIR SPYDER TESTS–BY DON MCKENNA

Run	*Condition*	*0–60 MPH*
1.	Stock	10.5 seconds
2.	Copper plug wires, 0.014 points	10.1 seconds
3.	Above, plus 1.2:1 fan pulley	9.6 seconds
4.	Above, plus wrapped exhaust	9.3 seconds
5.	Above, plus big paper air filter and 0.104 main jet (stock is 0.102)	9.1 seconds
6.	Above, except no air cleaner, 0.107 jet	8.9 seconds
7.	Above, plus velocity stack and 0.110 jet, no cleaner	9.1 seconds
8.	Above, without stack or cleaner, with 0.110 main jet	8.7 seconds
9.	Above, with muffler and weight removed	8.0 seconds
10.	Above, except with 2-barrel Rochester and muffler	8.2 seconds

Note: All runs except No. 9 made without removal of any weight, approximately ½ tank of gasoline, and driver only. Timed to corrected 60 MPH (63 on speedometer).

16

Spyders and Other Turbocharged Corvairs

When the Spyder "made the scene" in 1962, it was highly publicized. The Spyder, along with the Oldsmobile Jetfire, made GM the first automaker to explore turbocharging for the American consumer. Literally hundreds of articles were written about its engineering features, but the best presentation of all was seen by comparatively few people. This Society of Automotive Engineers Paper No. 531A, "The Corvair Turbo Supercharged Engine," was authored by two Chevrolet engineers, R. E. Thoreson and J. O. Bradford. Much of the performance information contained in this chapter was made possible because of data gleaned from their fine work.

Since the sixties turbocharged engines spread through the manufacturers as performance options. Besides GM, both Ford and Chrysler sold several turbo models. The push at that time was the same as for the Corvair; the performance of a V8 when you want it, and the economy of a smaller motor when you don't. Usage of turbos dropped off as manufacturers figured out ways to make the normally aspirated motors more fuel efficient. But tightening up of Corporate Average Fuel Economy (CAFE) standards has brought turbocharging of much smaller motors back in a big way. The application for day-to-day driving of turbocharged cars has been aided by modern fuel injection and inexpensive engine control electronics. You will see a further increase in turbocharged production. Although Corvairs were blown by various types of superchargers prior to April of 1962, the greatest number on the road today are turbos. Performance-improvement discussion has been restricted to that type.

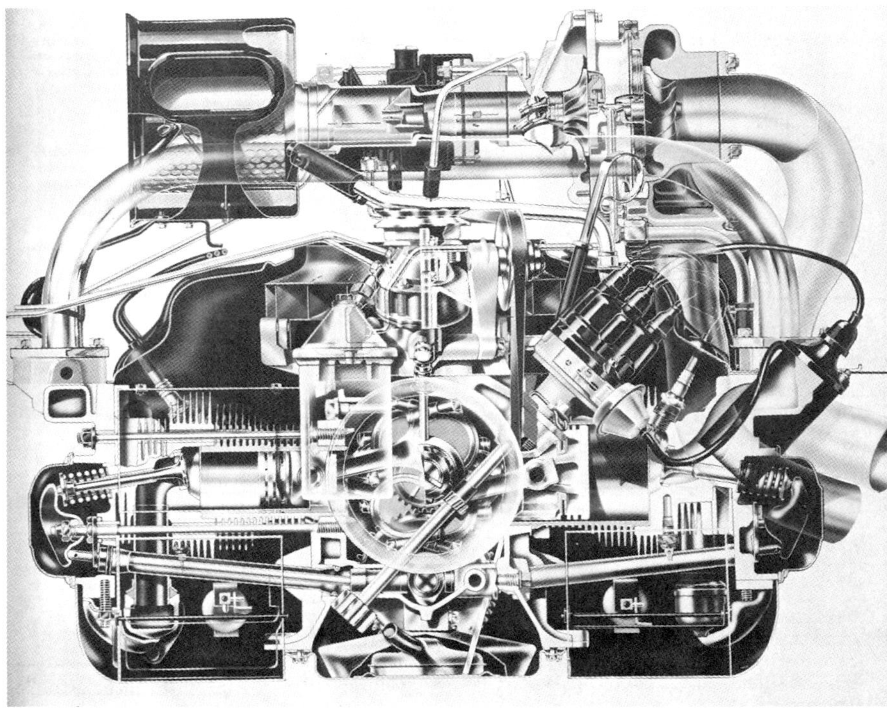

Ghosted view of '62 Spyder engine shows many interesting construction details. Changes for years '62-'65 are detailed in text.

General Description

An exhaust-driven supercharger forces air-fuel mixture into the intake manifold at greater-than-atmospheric pressure under certain load conditions to improve engine breathing and power output. The supercharger's "heart" is a balanced shaft with a turbine wheel at one end and a centrifugal impeller at the other. Each of these "wheels" has its own contoured housing. Hot exhaust gases, directed by a nozzle against the turbine blades, spin the shaft up to 80,000 RPM on a full-floating pressure-fed aluminum bushing bearing. The impeller draws air-fuel mixture from the carburetor into the compressor housing and passes it to the intake manifold under pressure. Supercharger output depends on load, throttle opening, exhaust-gas temperature, and engine RPM. Under heavy load, turbine speed increases due to increased exhaust flow and temperature, and the compressor supplies additional mixture under pressure to meet the demand, thereby increasing HP.

1962–63 Spyders differed from later models in that they were only 145 CID and lacked harmonic balancer type crankshaft pulley. Both can be updated to the later specification at rebuild time.

'65 Corsa engine gets 180 HP from an improved supercharger and bigger carburetor venturi. Supercharger has larger turbine passages and greater turbine-blade area to increase impeller speeds up to 20% over '62–'64 models. Compressor housing flow passages and blade are similarly increased to improve supercharger flow nearly 33% at peak speeds. Reduced flow restrictions increase volumetric efficiency by keeping mixture temperatures lower.

One of the Spyder's most-important design features is its pressure-retarded distributor assembly that backs off the spark setting as positive boost appears in the intake manifold. Advance is all-centrifugal, starting at 3,900 RPM and increasing to 12 crank degrees at 4,800 RPM. Retard of the spark depends on boost, with maximum retard of 6.5 crank degrees at 3.5 psi manifold pressure. Full-throttle advance is therefore the algebraic sum of the initial setting 24° BTDC plus the centrifugal advance 12°, minus the retard of 6.5°. Wide-open throttle at 4,800 RPM gives 29.5° (crank) advance. Pressure-retard specs differ for the various models. '62s showed 0° at 1 psi, 9° at 2 psi; '63s 0° at 2.5 psi, 8° at 4.5 psi; '64s 0° at 1.5 psi, 8° at 4.5 psi. These were the final specifications published for each model year. Only two distributor-pressure actuator assemblies are listed as replacements: one for 1962–64, and one for '65–'66 which has 12° retard at 4.5 psi boost.

By combining the pressure-retarded spark with 8:1 compression ratio, Chevrolet sidestepped a basketful of complexity, at the cost of some additional performance.

An important consideration in the adoption of a turbo-supercharger was normal aspiration under cruising conditions, which provides economical operation. Fuel-air mixture is then sucked through the idling supercharger as if it did not exist, the wheels just cruising along. When the throttle is opened wide and the exhaust system is heated up, a 10 psi (gauge) boost, or 1.7 atmospheres, can be realized at 3,500 RPM.

Thoreson and Bradford explained in their SAE Paper that an exhaust-driven

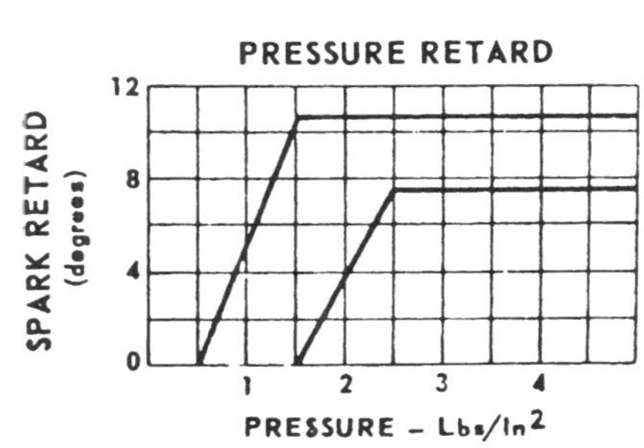

The spark is retarded when the manifold is pressurized, partially opposing centrifugal advance. Initial distributor setting is 24° BTDC.

turbo-supercharger was chosen for Corvair because of several other advantages: (1) exposed carburetion system for easy adjustments, (2) no noise or significant mechanical vibration, and (3) transverse mounting at the front of the engine used existing space most efficiently. Advantages obviously outweighed the comparatively minor bothers of boost-pressure lag due to turbine-acceleration time and high engine speed before positive manifold pressure is obtained (2,400 RPM at wide-open throttle).

Tom Keosababian's turbocharged Corvair—part of the Ralph Nader Racing Team—holds a 134.7 MPH 1966 Bonneville record. Standing-mile and ¼-mile speeds of 131 and 103 MPH, respectively, have been posted. Turbo whistles up 20 psig boost (10 psig in 1st gear) because of wrapped exhaust, big carburetor on IECO-type adapter, and milled case faces (0.025 to 0.035 inch) which reduced production tolerances. Water injection and 140 HP heads and camshaft also help. Freon sprayed through intercooler ensures cool mixture. Tom eventually reached 173 MPH in this car on propane.

How Do Spyder Engines Differ from Other Corvairs?

When the Spyder burst from its egg sac, the differences between standard and high-performance engines were extensive and numerous. However, by 1964, Spyder's important internal-engine changes were made standard in all Corvair engines: heavy-duty rod and main bearings, heavier connecting rod cross section, new piston rings and crankshaft, and new crankcase cover vent-baffle plate. Other changes, not all of which are obvious, relate only to the Spyder. These include a single Carter YH side-draft carburetor, new fuel lines, exhaust pipes, muffler, and heater ducts, oil feed and drain pipes for supercharger, 8:1 compression ratio, cylinder head temperature sender, and warning buzzer. Many of these changes are discussed extensively in other chapters of this book.

Additional changes included improvements in exhaust-valve materials and the inclusion of aluminum-bronze valve guides for exhausts (replacements are available from several sources). While the '62 Spyders shared a camshaft with the high-performance engines, '63s got a wild 324°-duration, 0.378-inch-lift "bump stick." In the '64 models, a cam is again shared, but this one has 298° timing with 0.390-inch lift. This same cam is also used in 1965–66.

All of the internal-engine changes should be made when adding a turbo-supercharger (or any other kind) to 1960–63 Corvair engines. Merely bolting on the turbo-supercharger without the improved internals and the all-important pressure-retarded distributor, would probably buy you a fair share of trouble. The low 8:1 compression ratio is also important when adding a blower.

Chevrolet Service News for April 1962 said, "The Supercharger unit should, therefore, never be removed from this special engine to be installed on another Corvair." This is still true for all 1960–63 engines other than Spyders, but the improved '64 engines should not suffer from bolting the turbo into position, provided the warnings about compression ratio and distributors are heeded.

Making Spyders Bite Harder

Improvements to the Spyder are much the same as those made to unblown Corvair engines. See the other chapters of the book. But, because of the peculiar requirements of turbo-supercharged engines, certain items require special emphasis and attention. These include exhaust, water injection, carburetion, ignition, and compression ratio.

Before starting to modify your 1962 or 1963 model, it would be a good idea to update it to the latest external specifications. These would include simple changes as detailed in the "Lubrication, Cooling, and Breathing" chapter: magnesium fan, revised blower bearing, larger oil cooler, and a spring loaded fan-belt idler. Spyders especially benefit from well-engineered accessory oil pans and rocker covers. When black anodized, these assist in eliminating heat from the oil and thereby reducing cylinder head temperatures as much as 50° F. Further details are in the "Lubrication, Cooling, and Breathing" chapter.

Exhaust Wrapping

The turbine-nozzle design creates back pressure in the exhaust system. Exhaust gases leaving the inlet nozzle accelerate to high velocity to spin the turbine. The amount the gases can expand (and the work they can perform) is dependent on their temperature at the nozzle. As exhaust temperature drops, the engine produces less HP, and blower acceleration lags way behind throttle opening. The stock system gives a constant 15-psi back pressure and 10-psi manifold boost at 3,600 RPM; and above (stock). Blower operation can be improved by (1) raising exhaust-gas temperature at the turbine inlet, and (2) by reducing restriction in the inlet path or carburetor side of the compressor. The extensive area of the exhaust manifolds, cross-over pipe and turbine-inlet pipe drastically reduces the temperature of the exhaust gases which leave the cylinder at approximately 1,400° F.

The first improvement to make to the Spyder improves low-RPM boost by reducing turbine lag and gives a net 15 HP improvement in power output. This is accomplished by wrapping the exhaust cross-over and turbine-inlet pipes with heat-proof insulating material. Covering these with aluminum foil reduces heat radiation still further—and protects the insulation from the elements and flying rocks. In the past, asbestos insulating paper and household freezer foil (0.001 inch thick) could be used, with water-glass solution used in forming the asbestos around the pipes. Or, an even better solution is provided through using an Otto Parts kit from Clark's. This kit included flexible ceramic jet-aircraft insulation and heavy-duty foil (0.003 inch) with special clamps for installation. But, as with any hop-up trick, there are always "jokers" or disadvantages. In this case, exhaust-pipe life is drastically shortened and a richer mixture is needed to keep exhaust-valve temperature from soaring due to the more frequently obtained back pressure. Failure to use a richer mixture would probably reduce exhaust-valve life. With occasional drag racing or other competition, you might have to replace the cross-over and turbine-inlet pipes once a year. The low cost and fantastic performance increase afforded by this modification (15 HP!) far outweigh any disadvantages. Vendors now have replacements, but not yet in stainless steel. Incidentally, an all-out engine would also be equipped with wrapped exhaust manifolds.

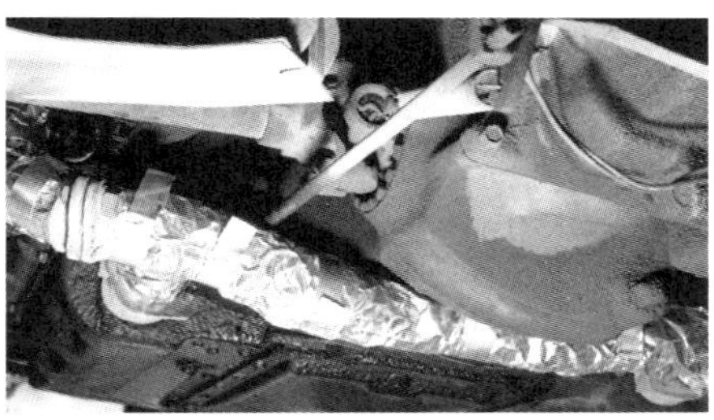

Photos of exhaust system on Glen Fotre's '63 Spyder illustrate special added flange to simplify engine removal, "lakes plug" header outlet and Otto Parts exhaust wrapping.

The Spyder exhaust system is an excellent place for application of Sperex VHT insulating enamel. You can use VHT in spray or brush-on form to paint the cast-iron exhaust manifolds, turbine, turbine pipes, and the cross-over. The high-dollar approach is to "Jet-Hot ceramic coat" these components. Jet-Hot cleans off the surfaces, inside and/or outside, then applies a coating and fires it in a furnace. This performs two tasks. It drastically lengthens the life of the pipe because the steel will not be burned away by the exhaust contact. It also adds to the heat insulation value of the pipe. When wrapped, it can help produce even more boost.

Dick Griffin, famous Spyder owner and holder of the 1962 G/Stock drag record, found that he got 0–60 MPH in 7.9 seconds and 16-second elapsed times for the 1/4 mile with three simple changes: (1) wrapped exhaust, (2) loose fan belt, and (3) 50 psi tire pressure.

Ignition

Spyder ignition has special advance characteristics and pressure retard. These cannot be changed in initial setting or advance curve unless other changes have been made—and unless you fully understand what you are doing. Spyder ignition systems must be kept in top shape for peak performance. Ignition is critical on these cars, especially those operated in competition! Improvements for racing include transistor packages, dual-coil distributors, and updated plug wires as described in the "Ignition" chapter. Decreasing the point gap will increase the dwell setting for improved coil saturation. This can result in reduction of point life. All such changes help to fire the dense mixtures created during supercharging. If you decide to upgrade the ignition system, keep the pressure retard unit in mind. It is unique to the Corvair. Any aftermarket systems must either be mechanically compatible with the GM pressure retard, or supply another method of retarding the spark under load. MSD makes a family of ignition products that plug together to perform many tasks. They have a boost-retard unit which could be paired with a magnetic pulse trigger inside the Corvair distributor, providing an adjustable clone of the factory's intent. At one point, Dale Manufacturing sold a modified early distributor, with a special advance/retard unit on it that did both, but it is no longer available. Vacuum advance on a turbo can really help the mileage—in non-boost situations. But you really need the retard under boost.

Plugs are usually the first suspects when a Spyder begins running "rough." Do not skimp on buying replacement plugs. If you have the AC-42-FF plugs, or the Champion L-12-Y, they will work. Both Denso and Robert Bosch both produce colder plugs for this application that can help projected-core nose plugs as discussed in the "Ignition" chapter. These recommendations are for normal, street driving. Racing a turbo requires the use of Champion L5 or colder plugs. Doug Roe points out that a careful watch on plug condition can tell you a lot about engine temperature and mixture strength. You'll note that all racing mechanics watch plugs closely.

Many folks have used water injection to counteract pinging on the turbo. Even the GM factory, on the Olds Jetfire, used a tank of "Turbo-Rocket

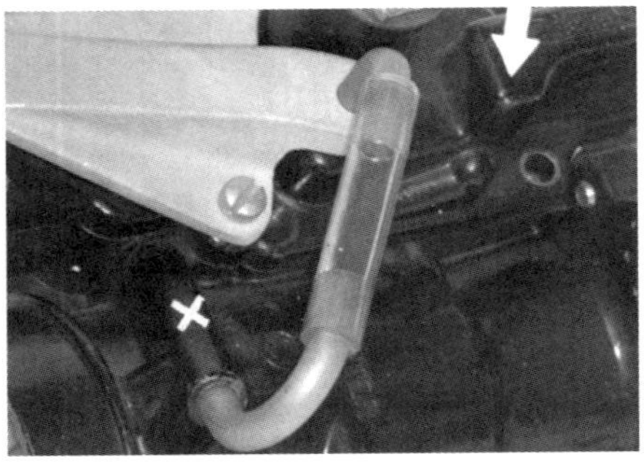

Anti-spill vent modification requires blocking stock vent at "X" and rerouting vent into top of carburetor bowl cover. Tube can also be installed into cover at point indicated by arrow. Installation here is with kit by Thomas.

Fluid", a gallon of which the dealer would happily sell you, to deal with detonation. There were several aftermarket kits available. The old HP Books title by Hugh MacInnes, *Turbochargers,* has a whole chapter on water injection. You have to be disciplined enough to keep the fluid level up, one of the failures of the Jetfire design.

Carburetion

Carburetion of the stock variety, the Carter YH, was near-perfect with the possible exception of flooding in corners, which can be eliminated by blocking the stock bowl vent and rerouting the air-horn pressure line to the top of the float bowl. Kits are available for this improvement. And, some of the earlier YH Carter carburetors were slightly rich on acceleration. Pump delivery can be reduced by scribing a groove in the top surface of the carburetor so a portion of the pump output bleeds back into the float bowl. Three separate main jets and metering rods have been supplied in the 1962–64 Spyders. If you have an early model and a fuel/air mixture gauge is available to use, you might find that additional performance could be added through changing to the latest jets. If a certain portion of the mixture range is not right, it may be necessary to modify the steps on the rod by carefully turning or sanding one or both down in a lathe. If the mixture is lean all the way through the operating range, then an even larger main jet can be created by drilling out the stock jet a few thousandths at a time. The Spyder is very responsive to careful tuning to get the mixture right and you can be sure that anything you do to alter the engine will affect the carburetor. This especially applies to removing the air cleaner, or substituting another type, or making changes in the exhaust system.

Dick Griffin's Spyder coupe won its class at the NHRA National Drags on two occasions. Full HP at the starting line was obtained by preheating the exhaust system by dragging the brakes as the car was brought to the line at full throttle in low gear.

The 1962–64 carburetor venturi is only 1-5/16 inches and this can be opened up by careful lathe work to get more boost more quickly and to add additional HP throughout the engine-speed range. Of course, the metering rod and main jet would need to be changed in conjunction with careful observations with a fuel/air gauge. As a starting point, a 1/8-inch diameter increase in venturi size requires a 0.105 main jet, with the 0.076 step on a '63 (2nd design) rod reduced to 0.062. Experimentation is essential to get it right! There are a couple of YH carb specialists still in business, Black Hawk Engineering in Colorado and Wolf Enterprises in Florida.

1965–66 carburetors have a 1-3/8-inch venturi, which is 1/16 inch larger than that supplied on the 1962–64 models. One of the performance tricks for the 164 CID motor is to use the exhaust wheel and housing from an early model, but with the compressor wheel, housing and carb from a late model. This combination will provide less throttle lag and more top end boost. And it all looks stock! Crown manufacturing used to supply two different special exhaust housings for the Corvair. The B-flow and the F-flow were designed to work with the appropriate exhaust wheel (B-flow '62–'64, F-flow '65–'66). Originally Crown had an adjustable stainless screw that the tuner could adjust to the right amount of boost. Eventually, they just made them at the highest boost setting and non-adjustable. When I asked Ted Trevor (of Crown) about this he told me everyone just cranked it all the way to maximum anyway.

The stock-carb venturi restricts boost. Dick Griffin found a 7-inch Hg restriction across the stock carburetor at 4,000 RPM. By increasing venturi area to 2.2 square inches he got 5 psi added boost and 34 more HP at 4,000. He used a Bendix-Stromberg WW downdraft 2-barrel carburetor No. 380499, originally on a GMC truck. With this installation he was able to chop his quarter-mile ET to 14.80 seconds, and raise the end-of-quarter speed to 93.6 MPH. Remember that this was with the 145-inch engine and the original turbo running in stock class.

Mike Burrett's turbocharger blows air through a Thomas manifold to pressurize two H Rochesters instead of sucking through a single Carter YH. Exciting performance is obtained.

Two-barrel carburetor kit by IECO is fine for '62–'63 models but must be used with discretion on '64–'66s. Water injection would be helpful with this installation.

Combustion chamber of '64–'65 turbo engine has no obvious quench area; is largest of Corvair chambers for 8:1 compression. 1965 turbos have same small valves used on the low HP Corvairs.

Car Life magazine reported on a similarly equipped test car as follows:

Performance measured	Car Life *Test Spyder (stock)*	*Griffin-modified Spyder (big carburetor)*
0–30 MPH, seconds	4.1	3.5
0–60 MPH, seconds	10.8	10.0
Speed, end of ¼ mile, MPH	80	84
Third gear boost, psi	6	11
Miles/gallon, hwy	20~23	19.2

Another development uses two stock Rochester H carburetors mounted at the usual unblown-Corvair position. Pressurized air/fuel mixture is fed to these carburetors (total venturi area is 1.57 square inches) through a generously proportioned cross-over pipe connected to the turbo outlet. See photo of blow-through engine owned by Mike Burrett of Fullerton, California.

The sewer-like cross-over pipe is 2-⅜-inch diameter, Master neoprene float valves are in the stock carburetors with 0.054 jets, and 27° BTDC is used for the initial ignition setting. Linkage, carburetors, and fuel lines are stock

Canadian Turbo

Dave Binnie has owned many Corvairs over the years and the turbos are his favorites. The original edition of this book inspired him to build another turbo. His '65 Corsa came from California and received a complete restoration at its new Canadian home. Originally white, it is now black with a black interior. The car had a 140 motor, it is now turbocharged via a unique blow-through system. The motor is bored .030" over with a stock 304 camshaft. 95 HP heads are used with some mild exhaust porting. The turbocharger is a factory turbo with e-flow compressor and .58 scroll on the turbine side. The only mods to the carbs are reinforced floats and rounded carb tops to permit the hose to be installed and sealed. The H-series lend themselves well to this application, but only the '60–'63 variety are used, due to later versions having vapor vents and enrichment valves. Ignition is stock 1110329 distributor with Crane electronics triggering a J+S Safegard unit. Many will laugh but he uses a stock 1960's Corvair coil and has never had a misfire! Dave is currently building a new, 140-based, blow-through engine using a Garrett turbo. Stay tuned. Dave Binnie lives in Ontario, Canada.

Corvair! Note that the $71 stock Spyder air cleaner/silencer is retained for its superb qualities. A balance tube from the manifold attached to the atmospheric side of the fuel pump maintains fuel pressures at the correct level. If constant high boost is used, be sure to keep an eye on the floats in the stock carbs. Eventually they will collapse from the pressure. The turbocharger in Mike's car goes to work at the top end of low gear—not third! A suitably lightened Corvair with this installation, plus water injection will run in the very high 90s at the drag strip.

Every type of side draft and almost every type of downdraft carburetor has been adapted to a Corvair turbocharger in the last 50 years. It is kind of a testimony to the ancient design of the YH and the unwillingness of most tuners to persevere with the adjusting of the Carter YH. The combination of the metering rods and jets can be tricky. SU side drafts and the Japanese Hitachi copy (from the 240/260/280Z cars) have been tried. These "variable venturi" styles should be perfect for the boost/no boost variable needs of the Turbo Corvairs, but the British carbs bring their own sets of issues with them. The Weber DCOE sidedraft carburetor has also been adapted to the Turbo. I ran one of these in 1972 on a built-up 140. When properly set up it was fast, but very thirsty. When modified for reasonable mileage, throttle lag ensued. LeVair Performance Products brought out a conversion for the 40 and 45 DCOE carbs, changing them to a progressive throttle opening. Much like the Holley 4-barrels, only the primary throttle plate is controlled by the linkage, the secondary is only opened when sufficient airflow through the primary will support the increased available venturi area of the second barrel. Another advantage of this progressive system is that the two throttles can be jetted quite differently, with the primary setup for semi-economical operation, and

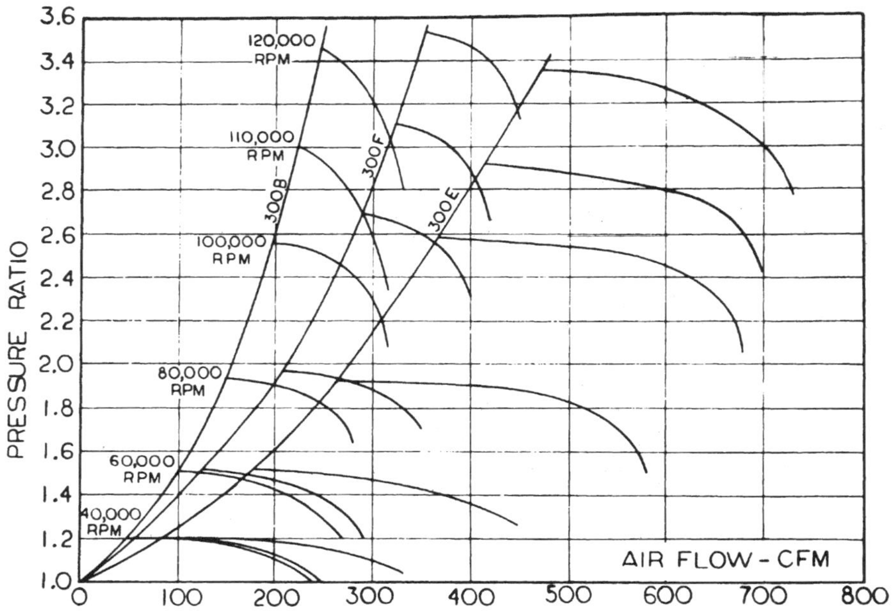

Composite chart shows compressor flow ranges for TRW Series 300B (used on 150 HP '62–'66 Spyders), 300F (used for 1965–66 180 HP Spyders), and 300E—available as a kit for use with 300B or 300F turbochargers. Note added flow of center unit (300° F) which explains how Corvairs got an additional 30 HP in the 1965-66 models over the same size engine in 1964.

Higher airflow capabilities for '65–'66 Spyders is offered by the "E" Flow compressor section on the left (compared with the stock compressor). The 3-inch diameter inlet requires a 2-¼-inch ID adapter to mount the carburetor. The complete kit includes a new housing, impeller, shaft seal ring and compressor, and two gaskets.

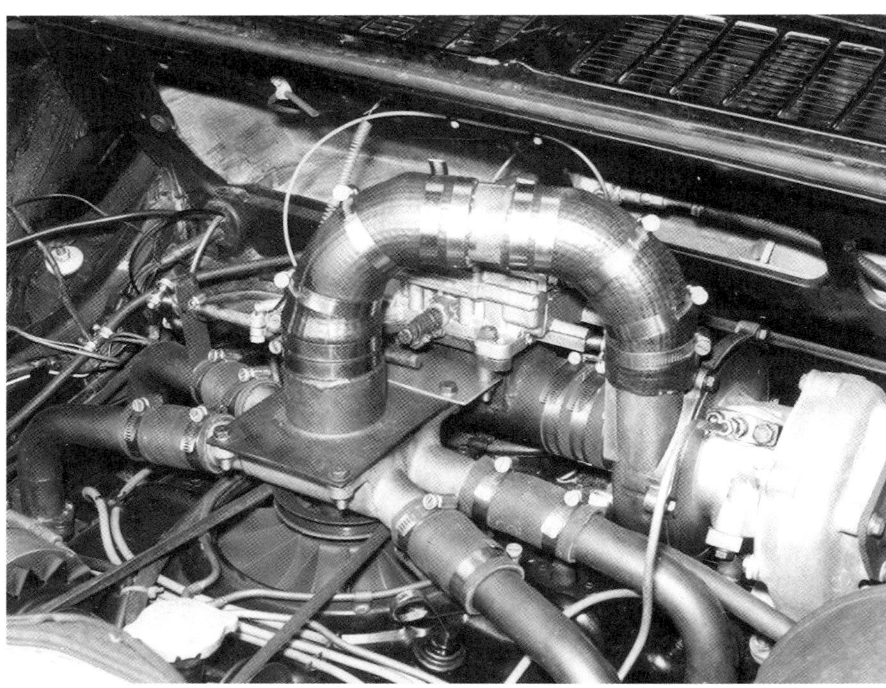

Engine room of the World's Fastest Corvair. AiResearch turbo-supercharger urges mixture from a Carter AFB carburetor. Water-injection components do not show in this photo. Relocation of muffler in engine compartment cleans up underside of chassis to reduce air drag.

This Weber two-barrel has been modified to open the secondary throttle plate based on airflow through the primary barrel. Note tube coming off the bottom of the carburetor.

the secondary throat setup for performance. A photo is shown.

Cylinder Heads & Compression Ratios

1962–63 Spyder cylinder heads are similar to the lowest HP unblown Corvairs. The really obvious differences are a recess for temperature sending thermister, different thread for temperature snap switch, aluminum bronze exhaust guides, and a supercharger oil drain (right head only). Combustion-chamber volume is the same as 80 HP models, giving 8:1 c.r. A radical change in 1964 Spyder heads (note photo) eliminated all obvious "squish area." This increased chamber volume to compensate for the additional 3.3 cubic inches displacement in each cylinder, thereby retaining 8:1 compression ratio.

If you choose to rework Spyder cylinder heads, be sure that you understand what you are doing if you increase the compression ratio past 8:1. Remember that cylinder boring and/or stroking increases displacement and raises the compression ratio. Spyder deck clearances need not be reduced as drastically as unblown engines, because the pressurized intake mixture provides turbulence in itself. A deck height of 0.060 inch should be adequate for the Spyder, and the step can be removed from the '62–'63 heads as described in the "Cylinder Heads" chapter. There is no deck or obvious "squish area" on the '64 and later Spyder/Corsa turbo heads. When modifying a '64–'66 Spyder/Corsa for competition (where rules permit) consider using the 140 HP big-valve heads. If valve-seat retention scares you, the 95 HP heads, like the 140 heads, are better than the original late turbo heads, the 95 HP heads have some squish area. A compression ratio as high as 10:1 can be used if the ignition settings are closely watched. Doug Roe ran at least this much compression on his engines with good success by reducing the total mechanical (initial) advance and increasing jet size to keep engine temperature down. Although a turbo oil drain can gravity feed a stock oil cooler placed in cool air, a pressure-fed cooler high in the airflow is a better bet. Oil must have a clear gravity-fed path out of the turbo cartridge to the head. The turbo oil seal is sensitive to oil buildup and leakage past here means smoking out the exhaust.

If you should need replacement aluminum-bronze exhaust guides, you will find that Chevrolet does not stock them. They preferred to sell you oversize valves or new heads. There are now many sources for the guides.

Besides the older Clark's EFI kit, the owner added an intercooled turbo. SDS electronics were used.

DCOE Weber adapted to the Corvair TRW turbo.

This beautiful engine also uses the progressive converted side-draft Weber. To the right is a wastegate with a manually adjustable control for the bypass.

Many different side-draft carburetors have replaced the original Carter YH. Shown here is a Solex variable-venturi unit.

Making It Handle

Rick Norris's Stinger at a Corsa track day event.

Improving Corvair handling is done to give you more fun from your everyday driving. Handling improvements will give you more constant pleasure than any amount of engine modifications. This chapter really should have preceded the engine improvement chapters as it is probably more important. The 1960–64 Corvairs (especially) respond eagerly when you bestow tender loving care and attention on the handling. The difference between a stock Corvair and one that has had a few minor suspension/brake/tire modifications is like night and day. You will not recognize it as the same car, but you will find that driving has suddenly become what you had always hoped that it might be. If you are competition minded, then these changes will increase your trophy-winning potential. The 1965 and later Corvair, although it is a great car for handling as it comes off of the showroom floor, can also be improved for an enormous increase in its fun-giving potential.

One of the men who contributed a great deal to this chapter of the book, Don McKenna, had his '63 Spyder coupe set up to the all-out modifications described in this book (for handling only), lacking only wide-base wheels to complete the competition package. Driving his car was a joy, even if you did feel the bumps more than you would in a standard Corvair. To my way of thinking, the ride can be compared to an MG-B or a Porsche. It is not as hard or stiff as the MG-A or TD, Triumph TR-3 or TR-4, Jaguar XK-120, or Sprite. By the same measurement, it is not as soft as a 190-SL Mercedes, Jaguar XKE, or a Corvette Stingray. You know that the car is going to do what you ask of it, which is more than you can say for many stock '60–'64 Corvairs.

Much has been written about improving Corvair handling. Unfortunately, many of the articles were written to sell a certain product. It has often been hard to determine where actuality was overshadowed by enthusiasm and sprinkled with misunderstandings as to what was actually occurring on the "road test." In getting detailed information to be able to write this chapter, Bill was most fortunate to be living in Northern California at the center of autocross competition (slalom, for those of you living east of the Rockies). The area was at least a year ahead of the rest of the United States as regards knowledge and understanding of setting up cars to handle over these tricky courses. I am referring to a general understanding by the competitors and not knowledge in the minds of a few gifted souls. Even Southern California took a back seat to these invaders from the San Francisco-Oakland Bay Area, as proved countless times in competition.

The situation was assisted by several very active clubs that posted impressive year-round autocross-race calendars. With a choice of several events each weekend, some competitors ran two events on a single weekend. This hotbed has spawned many formidable Corvair drivers.

Northern California rules may be different from those used in other parts of the country, as cars are classed according to their competition potential. Instead of running against other sedans, the Corvairs competed with all Alfas, Jaguar XK-120-140-150s, Mercedes 230-SL, Triumph TR 3 & 4, and Volvo P-1800, to name most of the class.

Three drivers were extremely helpful in providing details for this chapter and for the three-stage chassis-tuning table. One, Don McKenna, campaigned a 1963 Spyder coupe. He was second in class for 1963. In 2003, (forty years later!) Don was still racing, now driving a 1996 Corvette. The other gentlemen, owners of Corvairs, Porsches, and Volkswagens, are Glen Fotre and Erick Ressler. They constructed the fabulous Corvair powered VW sedan Rumplesteelskin, which often beat Cobras for top time of the day. With guidance from these men who spent their own cash for equipment and tires, truth and fiction were soon separated.

"Acceptable" riding and handling qualities mean totally different things to a driver, depending on his/her past experiences. In general, when I talk about making a Corvair "handle better," I mean making its response more predictable. By changing various items, the owner can make the car understeer or oversteer at will, and he can build "feel" into the car so that he can predict what will happen ahead of time. In a "handling" Corvair the driver feels as if he belongs to the car and is riding with it—rather than sitting on it and operating the controls as if on a skateboard! This secure feeling and improved cornering stability come at the expense of the pitching, floating ride which is so characteristic of the stock 1960–64 Corvairs. That the stock suspension is something less than perfect was pointed out by editors of *The Autocar* on January 6, 1961, as they tested a 1960 Corvair. They commented as follows: "It cannot be said that America's first car in large-scale production with all-round independent suspension shows any worthwhile gain in comfort or road-holding over its live axle rivals. Putting the engine at the back made independent rear wheel springing practically obligatory, but full advantage does not seem to have been taken of this. The ride is soft to the extent of sloppiness, with marked pitching over poor surfaces; it was considered that replacement of the standard dampers (shock absorbers) by units better suited to European conditions might well transform the suspension characteristics."

It is too bad that more Corvair buyers could not have read this road test instead of some of the whitewash treatments. Many people saw that the car had a rear engine and full independent suspension and jumped to the conclusion that Corvair was a sports car. You know, like a Porsche! The rear engine did not make the car a Porsche, VW, or grand-prix racer. The wheelbase, high roll center at the rear, high center of gravity, and swing axle all contributed to handling problems which were not helped by the ineffective shock absorbers as our English friends so aptly pointed out. Enthusiastic and/or careless drivers soon found themselves bargaining with the fender benders to repair damage from a spinout on a fast or slippery bend. Corvair was designed for riding and handling qualities equivalent to the 1959–60 full-sized Chevy passenger cars—not exactly a paragon of handling—and not for the type of driving which many owners thought it should provide. For this reason GM kept engineering more and more understeering tendencies into the car through 1964. After settling one suit out of court for $70,000, GM elected to protect the public from itself by changing to a rear suspension with decidedly more predictable characteristics. The result is seen in the fine handling of the stock 1965–69 Corvair.

Dave Edsinger's Prepared Class autocross car. Fastest time of day at Detroit convention, Dave road races the car at East Coast tracks.

The new Corvair's Corvette-type suspension is "something else" in any book. In general, however, many of the same improvements that are recommended for the '60–'64s will work on the '65–'69s because good handling requires similar modifications, regardless of the automobile being modified. It should also be mentioned that the modifications described in this chapter tend to stiffen the suspension so that you will feel even minor bumps in the road. But, ride smoothness is dependent on tire pressures, car speed, and the road surface, and you will be having so much fun with your transformed car that the bump portion will be quickly forgotten.

As in improving an engine, suspension and handling changes cost time and money. You can choose the degree of improvement which you require, working up to the various stages as you find your driving becoming that much more enjoyable. The parts that are needed are readily available. The table of suspension improvements should be very helpful to you as you set about to make your Corvair handle like a true sports car. The All-Out version, especially with Positraction (which I would recommend for any Corvair), is an exciting and completely predictable car. Understeering or oversteering can be selected to fit everyday needs by merely changing the toe-in at the rear. More toe-in moves the characteristics toward the understeer condition, which can be augmented by increasing the stiffness of the front anti-roll bar. Less toe-in tends to make the car oversteer. The steering characteristics, contrary to what you might hear from the bench-race crowd, are not influenced by the rearward weight bias as much as you might guess. In general, slow tight corners are negotiated quickest with an oversteering car, which is one with its rear-wheel adhesion reduced. Faster courses require understeer for the fastest times. An oversteering car, unless the driver has had a great amount of experience, will be too quick to spin as cornering speeds are increased. Also bear in mind that when you "lose" the Corvair in a corner, there is little chance of recovering from the spin, regardless of suspension modifications. Here is where that rear weight bias and 108-inch wheelbase work against all of your best efforts. Your 108-inch car will not, regardless of what you call it, ever attain cornering speeds on short tight courses like the 83-inch Porsche, 90-inch Cobra, or the 98-inch Stingray Corvettes, But, cut a section out of the center of your Corvair to make its wheelbase similar to the GM-Styling cars and you would have a top runner for sure, especially with the 1965–69 suspension. Mount the engine ahead of the axle and the combination would be hard to beat.

Weight Redistribution

Anything that can be done to reduce the rearward weight bias will assist in improving Corvair handling. Two items can be easily shifted to aid the situation. Both the battery and the spare tire can be relocated in the luggage compartment up front. The battery must be carefully secured in a suitable mounting. The spare is easier to tie down. If an event's rules do not require carrying a stated amount of gasoline and has no stipulations about having jacks and the spare tire on board, do not hesitate to jettison these weighty items. Reduced weight means increased acceleration and easier stopping, too. Make the rules work for you.

Heavy-Duty Suspension (HDS) Parts

Stock HDS components are a big help in getting improved handling from your '60–'64 Corvair. Sorting out the numbers from the parts book can be a frustrating task—especially in the coil-spring area—so to avoid driving the parts counterman out of his skull, I have included what may be some helpful numbers. These are all circa 1965, so availability is near zero today. Some ID tags remain, however, even on old parts!

Front springs for installing HDS onto 1960–64 Corvairs other than convertibles are Part No. 3817231. Front springs for the same task on '62–'64 converts are No. 3826939. Rear springs become more complicated due to the torque reaction of the engine that demands separate springs for right and left sides. Non-convertible '60–'63 use No. 3828552 on the right and No. 6257159 on the left. Convertible '62–'63 use No. 6257159 on the right and No. 3796318 on the left. HDS rear springs provide from ½° to 1-½° negative camber at static load. You can use them on 1964s if you remove the transverse rear leaf spring.

Three separate front anti-roll bars have been available as stock parts, but HDS options included only one of these, Part No. 3783523. It was inexpensive, $6 plus about $7 worth of parts to mount it and the cost of welding on necessary brackets. Another item in the HDS lineup was a heavy-duty shock absorber, used only at the rear. The early model Corvair has the same mounting pattern front and rear. The primary difference between the HDS and the regular shocks is in the valving as the diameters are exactly the same at one inch. Northern California autocrossers often use these HDS rear shocks at the front of their Corvairs, with even stiffer shocks at the rear. HDS shock for '62 is stamped 3171192 sells as Part No. 3166525. Part No. 3166566 for '63s is stamped 3178104. What is the difference? I do not know. Always keep these part numbers handy when cruising swap meets.

Rebound straps used on the rear axles of 1962 Corvairs with HDS are something to take off of your car, never an item to add. These lift the inside rear wheel still further during cornering, increasing the undesirable jacking-up effect which is discussed in the decambering details.

Decambering

Decambering, or setting the front and rear wheels at negative camber in the static or unloaded condition, is often done to improve Corvair handling. It is usually accomplished at the rear wheels, but the front wheels can also be decambered to improve the "sticking power" of the front end. The Sports Car Club of America allows this, even in stock classes. The process lowers the rear roll center to reduce weight transfer to the outside wheels in cornering. It also improves the cornering power available from the tires (as long as excessive amounts are not used) because the tire is flattened against the road in cornering to improve its contact patch or footprint. Dunlop Tires racing engineers pointed out the need for setting transverse-link-type rear suspensions (such as the '65–'69 Corvair) at about 2° negative camber for maximum cornering power.

Drawings are also shown from *The Motor* (British) to illustrate the jacking effect occurring during cornering with a swing-axle-equipped car such as the Corvair, VW, Porsche, etc. This is caused by the difference in cornering forces on the inner and outer tires. Cornering "jacks up" the roll center of the car above its static-loaded position, producing a change toward oversteer, and sometimes allowing the inside rear wheel to spin so that power is lost. To reduce this phenomenon, or keep its effects within reasonable limits, the enthusiast owner often sets the rear wheels of his '60–'64 Corvair at 3° to 5° negative camber in the static-load position. This accomplishes at least three things. It lowers the roll center and helps to keep the rear wheel from "jacking" so far that it loses contact with the ground. Then as "jacking" occurs in cornering, the wheels move toward neutral camber, or to a less-positive camber position than would occur with stock settings. Negative camber on the swing axle car also reduces ground clearance. The use of negative camber is an important part of chassis tuning.

Regardless of camber setting, if the rear wheels are allowed to hop, as caused by rough surfaces and/or overinflated tires, the "jacked-up" rear end will temporarily drop to the static-load position and cornering power at that instant will be reduced drastically.

Corvairs are not "pure" swing-axle cars, but use a combination of trailing

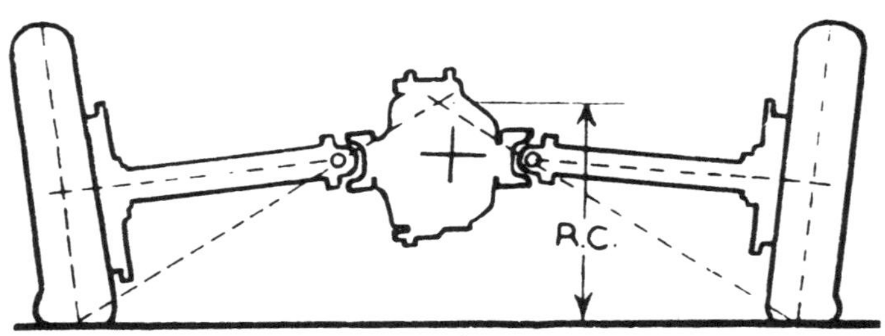

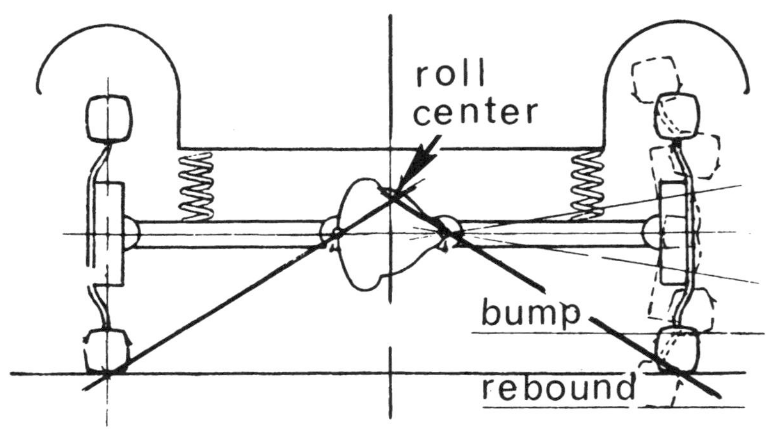

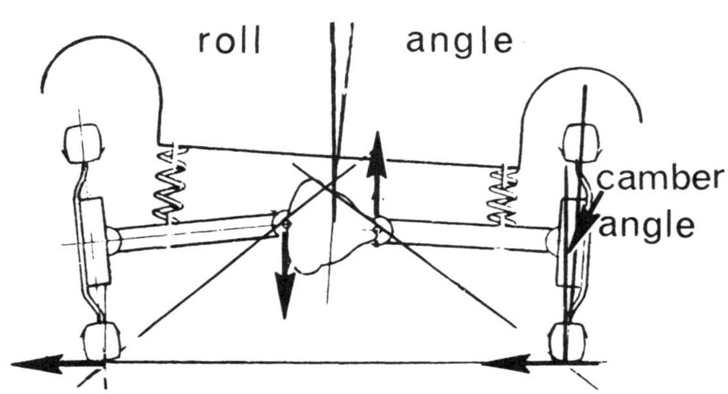

1960-64 REAR SUSPENSION

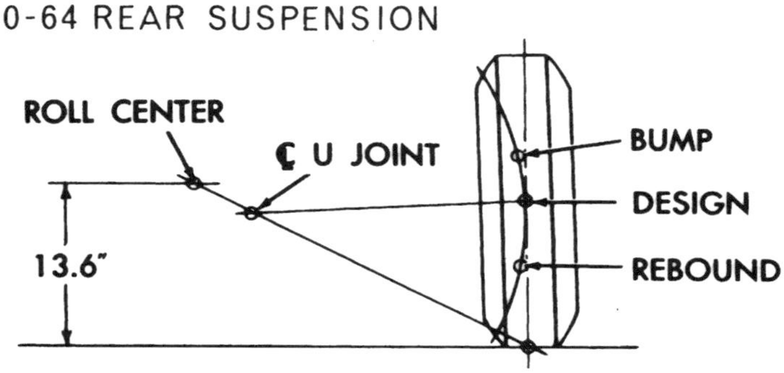

arm and swing axle on the '60–'64 models. The rear wheels swing in two arcs, one around the control arm pivot, and the other around the U-joint. This is similar to the idea used by Lagonda, Porsche, and VW. The net effect is to provide a slightly lower roll center than could be obtained with the true swing axle. Total camber change of the Corvair swing axle is 16.5° over 6.5-inch travel, including both bump and rebound (bounce).

1965–69 Corvair owners feel very secure in their cars knowing that sudden rear-end breakaway (slides) caused by changes to positive rear-wheel camber cannot occur. The '65–'69 setup includes inner and outer U-joints on each axle. Each wheel is controlled in camber by two arms, of which the axle is the upper control arm. The lower arm is longer than the axle and moves the wheel to a more negative camber in bump (jounce) conditions. If you set the rear wheels at 2° negative camber in the static-load position, the camber will progress toward an even more negative setting as the wheel moves up into the fender well.

There are disadvantages in using negative camber with each of the Corvair's rear-suspension systems: slightly uneven tire wear, and decreased ground clearance, the latter item especially referring to the '60–'64s. Decambering increases the toe-in (tending toward understeering), so the toe-in must be reset. Adjusting the tie rods provides

Severely decambered 1964 Corvair is regularly used for autocrossing. Recapped tire tread places maximum rubber on the ground. Note that top of wheel tilts inward.

the final settings on the front wheels, but, by removing shims at the front of '60–'64 transmissions—or resetting the forward location of the lateral arms on '65–'69s—to splay the wheels outward at their leading edges, you can get back to the desired toe-in setting. Or, you may want to use some toe-out, as this is the only way to offset excessive wheel scrub at decambered settings. Many autocrossers set the front to toe out for events, resetting it back to stock for street use. Toe-out makes the wheels tend to follow a straight path (tilt a wheel inward at the top and roll it to see what occurs). But, the setting for minimum scrub may not coincide with settings which provide the exact degree of under/oversteering which you find best for your own driving capabilities or for the type of course on which you most often run. Also, there must be a compromise between the 2° negative setting recommended by Dunlop for maximum traction in cornering and a setting which will get the rear end of '60–'64s low enough to offset jacking. Regardless of the experimentation that will be required to get the maximum benefit from your decambered Corvair, it is safe to say that any Corvair that is to be driven vigorously needs the benefits afforded by this simple modification.

You can decamber your '60–'63 Corvair by cutting one coil off of the upper end of each of the rear HDS springs. Installing uncut HDS springs at the rear merely gives ½° to 1-½° negative camber—not enough for hard cornering. Keep similar relative spring rates at front and rear by cutting a coil from each of the front springs as well. This lowers the front, but does not decamber it. Doug Roe used chopped 283 Chevelle springs at the front of his '60 Corvair racer. Non-HDS springs can be cut at front and rear, but these may not then be stiff enough to prevent the car from bottoming, even though chopping the spring stiffens its action. Always cut full turns from the springs. Heat the end of the spring and reshape to fit the spring seat if necessary. Cutting a full coil may lower or decamber the car too much, but the car can be shimmed back to the required height by installing "insulators" GM Part No. 6255922 (¼-inch rubber spacers).

1964 Corvairs used different spring rates with the rear coil springs accounting for 60% of the rear springing and a transverse leaf spring providing 40%. One or two coils can be cut from the rear springs, but longer high-tensile bolts must be installed to attach the transverse spring to the control arms. You can also remove the transverse spring and install shortened HDS springs.

1965 and later models have the camber of the wheel set independently from ride height. Wheels are decambered at the rear by adjusting an eccentric at the outer end of the lower control arm (see drawing). If this will not provide sufficient negative camber, and you are determined to use the original lower control arm, then shims can be added at the inner end of the lower arm to move the support bracket out from the transmission, although this can cause bolt insertion problems, especially for the vertical bolts. The '65–'69s also use a different method of rear-wheel toe adjustment as is illustrated. Total camber change on the 1965 and later rear wheels is only 8.5° over 7.5-inch travel (3.0-inch jounce, 4.5-inch rebound). The stock '65–'69 settings will give about 1.5 negative camber with four people in the car. The camber is approximately 5° negative at full jounce and 3.5° positive at full rebound. 1965–69 rear wheels do not stay parallel as has been described in some of the introductory articles. They cannot stay parallel because as the body rolls, the outside wheel must get more negative camber as it moves up into the body due to cornering forces. If the apparent negative camber increases, the true negative camber (in relation to the road) stays about the same. For this reason the lower control arm is longer than the upper arm (axle). Toe change on the rear wheels is controlled by a pair of links that stabilize the front of the trailing arm, by connecting it, via a link rod, to the front crossmember of the transmission. In mid-1965 those front control rods on the rear suspension, the part that bolts to the transmission crossmember, were dropped 1.75 inch at their inner ends (center of car) from parallel with the ground to a decidedly downward angle. This provides outside wheel toe-in on corners and, therefore, roll understeer. Early '65s can be converted to this later specification by installing new brackets, Part Nos. 3875055LH and 3875056RH. The original factory recommended specification was 1° negative rear camber. With the later brackets, the recommended setting is 1° positive. The bracket and camber changes cured "skitterishness" and hunting on wet and/or icy roads. This setup is recommended for general all around driving, but some negative camber is still important.

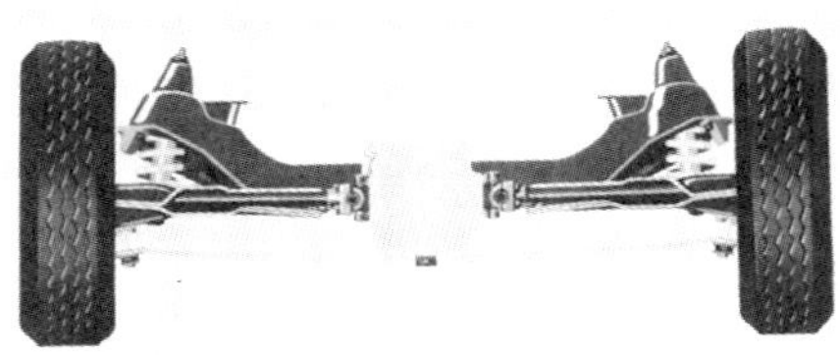

1964 rear suspension has lower leaf spring attached semi-rigidly to transmission case. This is not a camber compensator as it remains nearly neutral in roll conditions. It was added to increase understeer with a reduction in rear roll couple and the use of the heavy anti-roll bar at the front. Leaf spring carries 40% of the load, coils 60%.

Left-rear suspension on '64 model shows finned rear drums, added leaf spring. View is typical for '60-'63 models except for changes noted.

'65 and later vehicle height can be adjusted front and rear with spring insulators (shims), and lowering can be accomplished by cutting full coils from the coil springs. Heavy-duty springs are listed for both the front and rear of the '65–'69 models.

Decambering at the front is accomplished on 1960–64 models by adding shims at the inner attachment points of

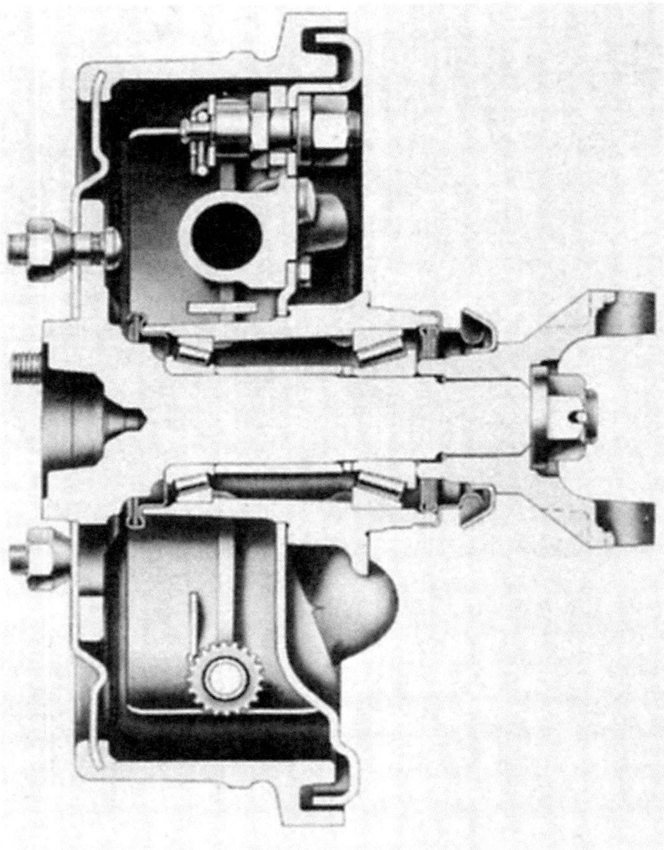

Cross section through rear-wheel hub carrier for 1965. Lower control arm attaches to trailing radius arm which also provides for spring and shock mountings.

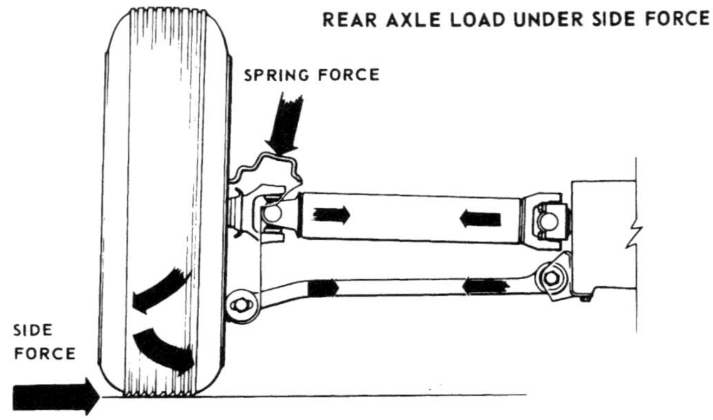

Typical view of right-front suspension showing compression strut for caster adjustment and anti-roll bar attachment. Anti-roll bar first introduced as ⅝-inch diameter on '62 heavy duty suspension, made standard at ¾-inch on all '64s, increased to 13/16-inch on '65s. Bracket is welded onto lower control arm in 1965 models.

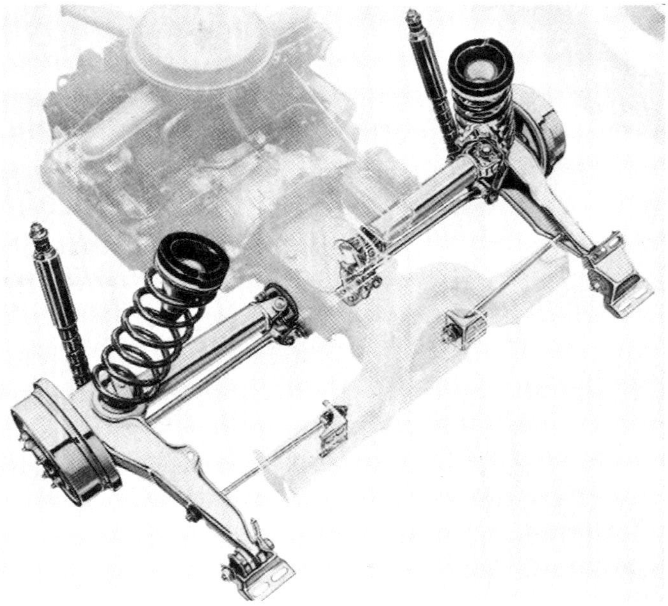

1965 rear suspension is race-bred, shares ancestry with Sting Ray, Cobra, and XKE Jaguar. Steady rods, toward front of radius arms, control side movement of arms which mount in big soft insulators for road-noise dampening. Camber adjustment is by eccentric at bent outer end of lower control rod. Axle is upper control rod.

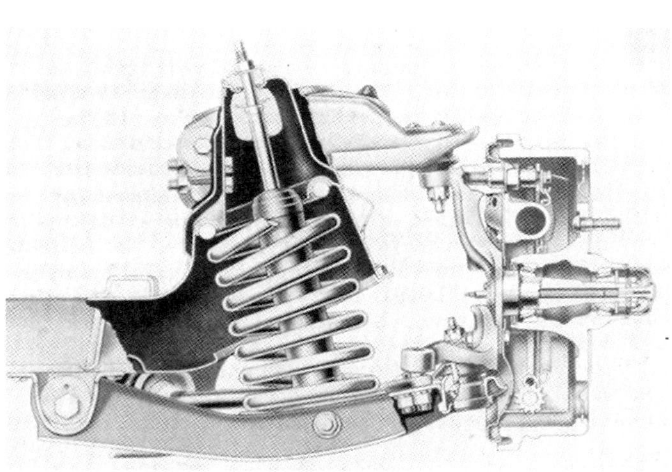

Cutaway section of left-front suspension assembly shows speedometer drive in hub introduced in 1965, shock through spring, ball joints, and compression strut which allows lower arm to be made much lighter and at the same time provides for caster adjustment. This view is typical for 1960-69 except for speedo drive, rubber bumper for lower control arm (moved to shock absorber in '65), and lower shock-absorber attachment method.

the upper control arms. Past a certain number of shims, longer attachment bolts are needed to accommodate the added shim thickness. 1965–69 Corvairs have front-wheel-camber adjusted with an eccentric at the inner pivot point of the lower control arm. If the eccentric cannot be made to give sufficient adjustment, then shims can be added as on the '60–'64 models. Approximately ¾ to 1-½° negative camber can be effectively employed at the front wheels to improve front cornering power, especially for competition use.

Front Wheel Caster

Added caster at the front wheels is sometimes helpful in reducing "plowing" of the front end when everything else seems to be working in good shape. Additional positive caster, created by pushing forward of the lower ball joint location, causes the front wheels to "lean over" in the direction of turning. Rubber in the caster strut, specified to perform a specific shock-reduction roll, allows enough travel to permit reasonable play in the steering. Clamping force of the caster strut on the late models can be increased by shortening the metal sleeve that spaces the clamping washers apart. There is also a new kit on the market that replaces the rubber in the caster strut with a small weldment and a large rod end. This will positively locate the lower ball joint. Be careful with increasing the caster by pushing the outboard end of the lower control arm forward. Forward movement of that point increases the wheelbase of the car. Eventually you may exceed a rule about changing the wheelbase. It doesn't take much to cause tire interference with the front of the fenderwell.

Anti-Roll Bars (Stabilizer or Anti-sway Bar)

It has long been understood by chassis builders that fixing a torsion bar across the chassis and coupling the ends through links to the suspension members (A-arms) resists cornering roll or lean without affecting up-and-down movements over normal bumps. This torsion bar must be fixed to a very rigid portion of the chassis, but left free to rotate in its bearings. If the opposing wheels are raised and lowered in unison, the anti-roll bar has no effect. Raising only one wheel twists the torsion bar and its natural springiness resists the twisting movement. As the car leans in a corner, the effect is to raise the outer wheel and lower the inner, and this movement is resisted by the bar, transferring weight to the outer wheel. This adds understeering characteristics at the front of the car. The stiffer the bar, the more weight transferance, and the more the front wheel on the outside will "wash-out" or "plow." There is also a diagonal couple from the outside front wheel to the inside rear wheel so the anti-roll bar also aids in keeping the inside rear wheel on the ground—which is helpful regardless of whether or not you have Positraction.

The original Corvair front suspension design included an anti-sway bar. According to history, the bar was removed in the production design by the accountants, over the protestations of the engineers, to keep the low price for the car in the face of increased R&D costs. It wasn't the first time or the last that accountants had strong-armed the engineers to keep costs down. That inclusion, then removal, came out in the courtroom discussions regarding testing of handling on the first Corvairs. The first two years of production Corvairs did not have a front sway bar.

Anti-roll bars were first introduced for Corvairs by EMPI, subsequently by Eelco, and then the 1962 HDS-option kit was made available around April of that year. This ⅝-inch diameter bar, also optional for '63, is Part No. 3783523. Every 1964 had a ¾-inch bar and all '65–'69s share a 13⁄16-inch bar. Stock anti-roll bar parts cost about $13, but the necessary brackets must be welded into place. Today, it is much easier to just swap in the front crossmember you want, brackets already installed. You can swap the above bars between models to "tune" the amount of weight transfer you want. EMPI- and Eelco-made anti-roll bars attach without welding. EMPI bars work "full time" as the mounting points are close to the outer edges of the control arms to which they attach. This makes the bar work at smaller wheel deflections

Don Eichstadt shown in a fast sweeping bend at track near Detroit, leading Cobra-Weber ponycar. Don provided many ideas which have been incorporated in this handling chapter.

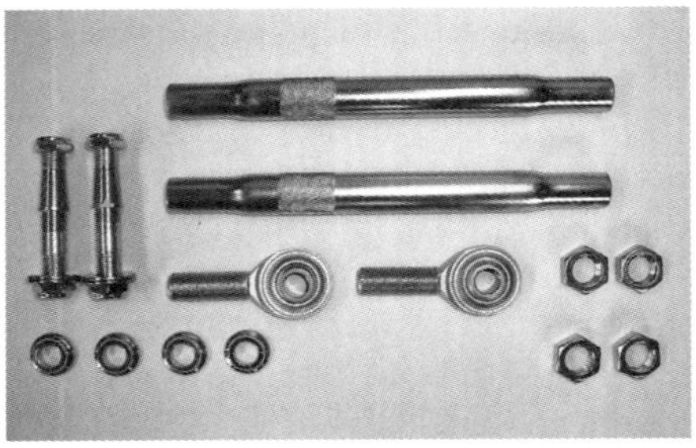

Bump steer adjusting tie rod kit as received.

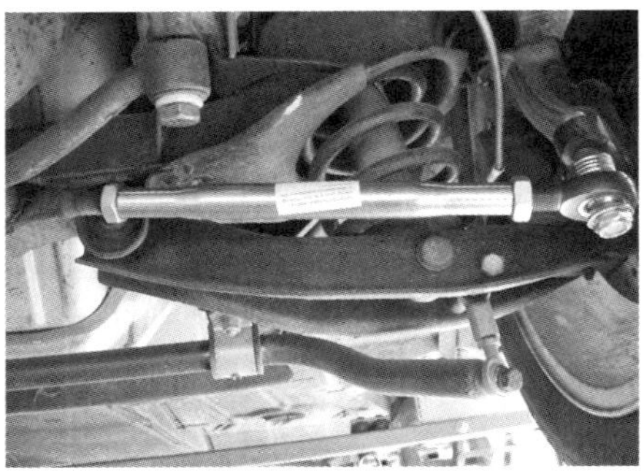

Bump steer adjusting tie rod kit, installed on a late model. The take-apart design allows the rod end to be shimmed up or down to change the bump-steer characteristics of the front suspension.

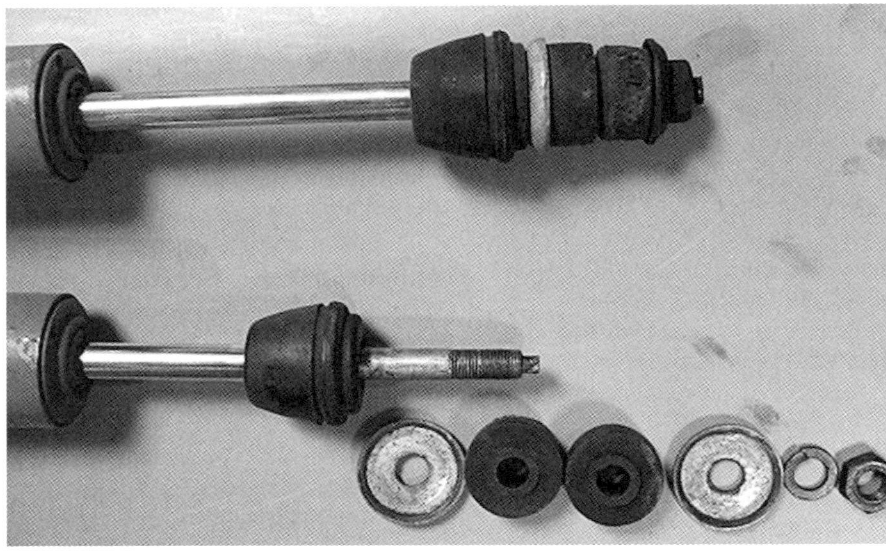

Koni Shocks come with a travel-limiting bumper. This bumper is similar in design to the original design used on late-model Corvairs. The Corvair bumper was much larger.

than the stock bars. And in some situations, it is advisable to use both a special and a stock bar, as detailed in the chart. Experimentation is quite often needed to get the exact amount of front anti-roll bar action that will be best for the use being given to the car. For instance a '62–'63 HDS-equipped car, still tending to oversteer, seems to benefit from installing a '64 bar.

1965–69 models have a tendency to "plow" or understeer from the factory. Chevy was really working on reducing the oversteer of the early models. The late models can be changed in the direction of oversteer by fitting an earlier, lighter weight bar (smaller diameter). In a class where aftermarket sway bars were not allowed, I swapped a '64 bar onto a '65 and the car worked much better in tight slaloms.

Crown Manufacturing had offered several sway bars, both front and rear for the late model, when they were selling the V8 kits. Many were sold for the regular Corvairs as well. Their kits came with the brackets needed to replace the stock front chassis mount and add a tip mount to the outer point of the lower control arm, where the caster strut attached. Like any street piece at the time, Crown used rubber both at the bar mount and the tip mount. At the rear, Crown added two brackets that bolted on, but worked much better when welded on, and required only a couple of small holes in the trailing arms. Only minor modifications to the heater boxes are required to install the rear sway bar in a Corvair-engined '65-'69. No mods are required for installation in the Corv-8. There have been a few copies of the Crown units made, IECO and others sold them. There was also an Addco rear bar produced. Crown fronts and rears came in three sizes, originally, ¾", ⅞" and 1", but 70% were the ⅞" bar. Clark's still sells a version of the Crown front bar. They also sell an adjustable rear bar supplied by Vair Fast Engineering. That company is working on a line of tubular front sway bars as well. Although most of the sway bars came with rubber mounts, there are duplicate hard Urethane pieces available which

HANDLING KIT PERFORMANCE TEST TABLE

	5.5 × 13 wheels *6.50 × 13 tires*	*8 × 15 wheels* *10.35 × 15 tires*
Corvair Corsa coupe with Crown's GM sintered-iron brakes, 140 HP engine	40.4 seconds (stock suspension)[1] 39.3 seconds (kit installed)[2]	39.0 seconds (stock suspension)[3] 36.5 seconds (kit installed)
Corv-8 with Crown suspension kit, 350 HP engine. Non-GM metallic brakes increased time by 1.5 seconds by driver estimate.	37.6 seconds	33.0 seconds

(1) Pronounced roll and understeer on sharp turns
(2) Shows that stock suspension camber changes prevented using big rubber to best advantage
(3) Demonstrates need for wide-tread tires for best handling, even with handling kit installed

A then-new '62 with HDS and Goodyear T4 6.50-13 Bluestreaks at Waterford, Michigan Spring Sprints—cutting inside of VW/Porsche.

Second photo shows '65 in race near Detroit.

Don Eichstadt of Berkeley, Michigan, owned and drove both of these machines.

will stiffen up the whole system of the bar, making them more effective.

Shock Absorbers

Several brands of "just-for-Corvair" shock absorbers are available, most of them at "dear" prices. Unfortunately, the small access hole for installing the shocks on 1960–64 models prevented the designers from building the best possible shock for the car. Thus, they have been frustrated into supplying compromise units to fit through the available openings. Such compromise shocks are generally too small for vigorous driving. Racers are not so fussy about using shocks which demand special installation techniques. Hence Northern California autocrossers in general have become addicted to the use of Gabriel AdjustOMatic and Silver "E" shock absorbers. Most of the more successful '60–'64 Corvairs use units cataloged to fit the front of 1957 Chevrolets, especially at the rear of the Corvairs. Units from other cars might "fit" as well, provided that their extended lengths and mounting methods were similar to stock dimensions. Otherwise, a fully extended rear suspension might tear the shock apart at maximum travel. It is not unusual to find successful autocross Corvairs with HDS rear shocks mounted at the front, with Gabriel Silver "Es" at the rear. No modification is required to use the factory HDS shocks in the front position on your '60–'64 Corvair.

In 2003, Koni reintroduced a specific shock available for the Corvair. First on their list of reproductions was the '65–'69 shock, but they have announced shocks for the '60–'64 as well.

Replace the rear shocks on '60–'64 models when the springs are out, especially when installing larger shocks than will fit through the hole in the lower control arm. The big shocks must be installed into the topside of the control arm, then the special or shortened spring set over the shock and pushed into the upper "hat" section. The lower arm can then be jacked up and the shock piston extended to permit attaching the holding hardware at the top of the shock rod. 1965 rear shocks were not contained within the coil spring, so replacement is easy.

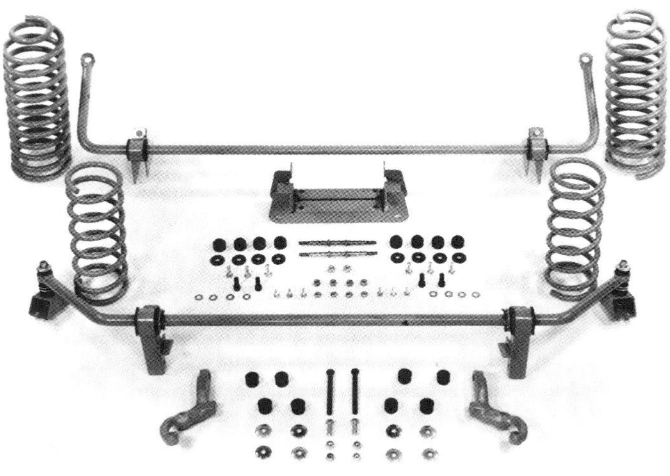

Handling kit for the '65-'69 Corvairs adds a rear sway bar, changes rear-suspension inner-pivot points for the lower struts, adds quick steering arms, and revises spring rates. Wider tires and metallic brake shoes are essential to take advantage of the improved handling afforded by this Crown kit.

Rear of '65-'69 with handling kit installed in a Corv-8. Note revised lower strut inner pivot points, changed springs, and rear sway bar.

'65-'69 front suspension removed from car to show mounting of handling kit components with revised pickup and mounting points for sway bar.

Lamar Engineering spoiler for '65-'69 Corvairs increases stability and cooling. Spoiler dams air in front of it, causing increased pressure over the entire rear deck. This down force stabilizes the car and improves cornering.

Don Eichstadt suggests trying the Gabriel Adjustable E Shocks for Chevrolet trucks, for use on the rear of '65–'69 Corvairs. The problem with all shocks is that even if they fit, with the right mounts, the right travel, the right lengths, they still may be the wrong shock for your car. At least the adjustable shock gives you a chance to adjust the rebound.

Quick Steering

The standard steering ratio for the Corvair never changed. Because the Corvair had no power steering, it was not really needed in most cases, the ratio had to be slow enough to allow underpowered drivers to still parallel park the car. The resulting steering ratio (20:1) was too slow for spirited driving. One of the first accessories on the market for the Corvair was a set of faster-ratio steering arms. The steering arms mount out at the front spindles. Shortening the arm length from the spindle to the tie rod, allowed the same movement of the steering wheel to cause the front wheels to turn a greater distance, effectively "quickening" the ratio. Shorter steering arms were made by several sources in the '60s and '70s. IECO, Eelco, Fitch, and others marketed them. Most were made of cast steel. They were and are

Four different driver's-side steering arms. From the bottom up: factory quick steering, factory regular steering, modified long arm to duplicate factory short, and aftermarket really short arms. It is easy to see what a difference the length makes in handling.

The Flaming River quick-ratio box bolts into any year Corvair car, some years require an adapter to mate with the steering column.

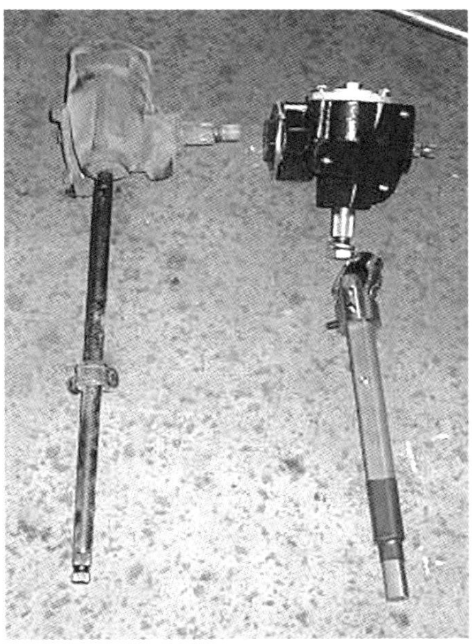

Original 1966 box is shown on left. Flaming River replacement box on right uses a collapsible center connector to install into the late 1965 and all 1966 Corvairs.

a valuable accessory for the Corvair. These reduce the number of steering-wheel turns (lock-to-lock) and are helpful in improving the responsiveness of all Corvairs not already equipped with the factory quick steering ('65–'69). The advertised claim of under three turns is seldom achieved with the arms—it is usually more like 3.7 turns—but this is still a marked improvement over the stock 4.6 turns. The arms can interfere with the frame, which can be mashed back out of the way for clearance, and have a couple of undesirable side effects, the diameter for a U-turn is increased, because the wheels don't turn as far, and the stock Ackerman effect is slightly altered. This is only noticeable in tight turns as tire scrub. Steering effort is increased, of course. The factory quick steering option included a redesigned steering box with a reduced steering ratio inside the box—dropping from the original 20:1 ratio to a 16:1 ratio—and shorter forged arms. These were not as short as the aftermarket arms, but are free of the undesirable side effects, too. The factory quick boxes have long been sought after by racers and have become hard to find. Reproduction quick-ratio steering boxes entered the market several years after the factory quick boxes disappeared off Chevy's shelves. They were marketed by Flaming River Industries of Beria, Ohio. By themselves, they can make a noticeable improvement in steering response, about 60% of the way to the factory ratio. To duplicate the overall factory ratio, shorter arms must also be installed, but the faster box may be all you need. Either the factory box or the Flaming River box may be installed without having to re-align the front end. The Flaming River boxes were last produced in 2017, and are not currently available. I cannot recommend the combination of a quick box and the aftermarket fast arms for anything but absolute racing use. Besides the tremendous increase in steering effort, worsened by today's sticky tires, the car's handling becomes "twitchy," almost nervous on the street. Richard Finch, in his book *How to Keep Your Corvair Alive*, recommended shortening the factory forged arms to quicken their response. With a couple of cautions, I can recom¬mend this. First, only a qualified (that means certified) welder should do the prep and weld on these arms. If you have a failure here, something bad is likely to happen shortly after. Second, after the welding, a qualified metallurgy shop should check the arms for any cracks in the weld area. Only then should you use them.

Steering Damper

A seldom-seen Corvair accessory should be given serious consideration: the steering damper. These reduce driving effort over rough surfaces, but should be disconnected for most racing use. These parts were available from Sears, J. C. Whitney, and Clark's.

Drag Racing Suspension

Numerous articles written about the Corvette-bred suspension of the '65 and later Corvairs have extolled the virtues of this only U.S.-made car ever to be developed and sold for less than $3,000 with a race car-like suspension. It took the development program for Crown Manufacturing's Corv-8 to uncover a serious suspension fault which hampers the Corvair from being the best possible high-performance chassis for drag racing. This fault is a severe camber change at the rear which prevents keeping larger tires on the ground for maximum traction in drag racing starts. The revised lower strut inner pivot point bracket developed for '65-'69 Corvairs is shown in the accompanying photos on page 124 and its performance depicted in the table. It minimizes rear camber change throughout rear-wheel travel and makes the front roll bar work effectively. The kit can be quickly installed using only hand tools. The bracket also acts as a protective device (aka skid plate) for lowered cars. The relocated inner mount geometry is specific for drag racing, to keep the wider tire flat on the ground. I do not recommend this geometry for the autocross or track car.

SUSPENSION TUNING TABLE

ROAD	DUAL PURPOSE	ALL OUT
Springs, Front		
Stock springs or the HDS replacements. These will lower the car from original height.	Cut one coil off the HDS springs. Cutting the height lowers the car but raises the spring rate.	Look for replacement springs of the same height and coil numbers. Wire thickness will determine rate. Convertible HDS springs can have two coils cut. Don't go too stiff on fronts.
Springs, Rear		
Stock HDS spring will lower the car slightly. Early models can go to '64 coils and lengthen the outer links to the suspension of the transverse leaf spring, to slightly lower the car.	Cut a coil on the spring to lower the rear a bit. Be careful on early models. Raising and lowering is your only camber adjustment. Late models have an independent method.	For autocross or track, it is hard to go too stiff on a late model. Pigtail springs are available for a perfect fit and a rock-like ride.
Anti-Roll Bar, Front		
Adding a roll bar for the early models may be most easily done by swapping in a complete '64 front crossmember. The stock '64 bar will help. The '65–'69 bar can be installed for a little more stiffness. Addco makes a ⅞-inch front bar for the early model. For a '65–'69, swapping to the '64 bar will increase oversteer. Handy in tight autocrosses.	The early model can swap in a '65-up front crossmember. This installs with the thicker '65 bar and allows a Crown front bar to be installed as well. (It also includes the bigger '65 brakes!) Crown or other front bars, IECO/Addco/etc. are enough for a dual-use car. Crown installation should be done carefully to maintain alignment.	Tubular front and rear bars of various OD and wall thicknesses are available for the late models. Go stiff on the front and tune the adjustable rear to get the balance you want. You can stiffen the rear for autocross oversteer and soften the rear for predictable track balance.
Anti-Roll Bar, Rear		
For most street use no rear bar is needed for any Corvair, early or late.	Addco does make a rear bar for the early or late model. The smaller original Crown style rear bar, when combined with a stiffer front bar can produce a good F/R balance.	Newer, stiffer rear bars are now available for the late models. Most still use the Crown mounts to the chassis, but offer better connection points to the suspension. Again go stiff on the front and adjustable on the rear.
Steering		
Stock steering on the Corvair was slow. A common accessory on an early model was a pair of quick steering arms, despite some of the ill-effects of the swap, including increased turning radius. These are still the inexpensive way to go.	Factory quick steering was offered late in the '65 model year. Flaming River made a replacement '60–'63 box that bolts into the '60–'63. It is almost a bolt-in for the '67–'69 as well. Replacing your box with the Flaming River unit is the best way to quicken up your steering response.	Many Corvair race cars have gone to rack-and-pinion steering. This works if the gas tank has been replaced, the pinion goes through the edge of the stock tank. If you have a stock fuel tank, Flaming River previously produced a box with shortened stock arms that works great.
Shock Absorbers		
Early models will work well with heavy duty shocks all around. (All four are the same) If you want to spend the money, put on Koni shocks after setting them to soft. They are available for lates as well. For less money, Clark's has Gabriels and others. A 1957 Chevy front shock will fit the early Corvair.	Set the Konis to medium on the rear. Other shocks might benefit from stiffer settings. Gabriel Striders are also adjustable. As you go to stiffer springs, you should stiffen the shock, because the "shock absorber" is really damping the movement of the spring. Heavier spring=heavier damping.	There are several double-adjustable coil over shocks that can be fitted to the Corvair if you wish to spend the money. The Koni shock for the Corvair can be re-valved at the Koni facility in the US if you actually need a stiffer rear shock. If you have adjustable shocks now, try them at full hard and see if it makes a difference first.
Wheels		
Finding any performance tire in original Corvair 13-inch sizes is almost impossible. Why bother? Both earlies and lates have lots of 14-inch and 15-inch wheels that will just bolt on. For road use: 14×6 wheel for an early, a 15×7 wheel for a late.	Same wheel selection as for road use. Another option, have a set of wheels for normal street driving, another set with stickier tires for "performance driving," like having winter/summer tires, but more fun.	There are still many 13-inch race tires available, because many small cars are still raced. An early model can fit a 7-inch-wide wheel in the wheelwell, at least for racing, late models, a 15×8 is easy. A little close on the front, depending on offset.
Tires		
Earlies work great on a 185/70-14, inexpensive size and easily available. Lates a 185/65-15 or a 205/55-15 for a more aggressive look.	Earlies can gain a little torque with a 195/60-14. It is also available in softer compounds. Lates can use a 225/50-15 or a 225/45-15 in sticky treads.	Several folks have stuffed a 10-inch-wide, 15-inch wheel on the back of a late model. Some wheelwell surgery will be required
	Maintain front/rear pressure differential	
Alignment		
Once the car is at a ride height you want, you have some autonomy to set the alignment how you want it. Rear camber on the early is determined by the ride height settings.	Road use and Dual purpose are the same: caster 3–4 degrees Front early camber 1.5 degrees negative Rear early camber 2-3 degrees negative Front late camber 1.5 degrees negative Rear late camber 2.5 degrees negative Toe settings: ¼ to ⅜ inch toe-in front Toe settings: ⅛ to ¼ inch toe-in rear	Front Camber at 2.5 degrees negative—Earlies/Lates Rear Camber at 4 degrees negative—Earlies Rear Camber at 2.5 degrees negative—Lates Toe can be set to zero for autocross and set back to normal toe-in for street driving.

Brakes, Wheels, and Tires

Stock, non-metallic 1960-64 Corvair brakes are totally inadequate for serious hard driving. Almost all U.S.-made cars of the era suffered from the same lack, so this is not an exclusive black eye for the Corvair, Mr. Nader. Wider shoes on the '65–'69 models, used in conjunction with 9.5-inch-diameter drums, increased the braking area to 169 square inches from the 126 square inches available with the narrower shoes and 9-inch-diameter drums of the '60–'64 models, and made a solid improvement. The elimination of asbestos as a material for brakes linings has definitely impacted the replacement shoe market for older cars. The big brake parts suppliers, such as Bendix, Raybestos, and Wagner have developed non-asbestos compounds that may perform as well as the originals. If you are going to replace original brake linings with new, but original-type linings, you would do well to purchase the shoes from a Corvair parts supplier. They are more likely to fit correctly the first time. If you are uncomfortable working on your own brake system, search out the oldest brake shop in your area and look for the oldest technician. He is more likely to know how your drum brakes should properly function. They might let you supply your own parts, they may require it.

Metallic Linings (Choice #1)

If you like to stop every time that you step on the pedal, then you have to make a choice on upgrading your brake system. You can install metallic brake linings with your existing drum system. These linings work better as they are heated up in hard use, continuing to provide stop after stop of safe, controllable braking. *Sports Car Graphic* for January '63 compared Corvair stock lining with replacement metallic linings. Their tests showed that stock Corvair brakes were "going out of the window" by the third stop, After only eight stops from 55 MPH, the brakes could not even hold the car against its idling engine with the Powerglide in Drive range. Metallic brakes also offer exceptional wearing capabilities. I know of one Corvair that has traveled over 75,000 miles, including much severe competition use, and the original factory Heavy Duty Service (HDS) metal brake linings are still intact. Metallic (sintered-iron) brakes were offered as HDS options on 1962 and 1963 Corvairs, but the factory parts are no longer available. The metallic-brake conversion kit for '64s could be used to install these brakes on all '60–'64 models, Part No. 3851093 identifies the kit, if you can find it. 1-inch diameter brake cylinders should be installed on the front when installing the kit, Part Nos. 5460812 (right) and 5460813 (left), but the smaller cylinders can be bored to the new size and fitted with 1-inch internals. Matching master cylinder, Part No. 5464275 can be installed if you need a new master cylinder, but it is not absolutely essential. Kevlar linings have also become available for the early drum brakes.

Increasing the front wheel-cylinder size, regardless of the lining type used, tends to reduce rear-wheel lockup and sliding. This is often a problem when

Stock metallic brakes are ceramic/metallic pads bonded to the shoes. Efficiency increases with temperature: the harder you use them the better they stop!

only the driver is in the car. Although backing off the brake-lining adjustments in the rear can help for slalom and road-race competition, the only real cure is to install a proportioning valve from the disc-braked Camaro, Part No. 3905525, Valve Assembly, Brake-Pressure Regulator. There are also adjustable valves available from the racing parts places. These are usually mounted within reach of the driver, in the line to the rear of the car. The valves limit the pressure to the rear brakes letting the fronts do more of the work.

When Chevy used parts common to the Chevelle to update the Corvair brakes for 1965, they swapped the linings front to rear from their Chevelle locations. The Corvair, having a rearward weight bias, needed the wider shoes on the rear. The front-heavy Chevelle needed them on the front. Chevy provided metallic brake parts for the '65–'69 Corvairs, but hid them in the Chevelle parts bins. The "factory metallics," as they are called, are highly prized by vintage Yenko Stinger road racers, because the stopping ability of that material has never been duplicated, at least not in the aftermarket for Corvair brake linings. The Vintage classes limit the racers to the drum brakes originally installed. Aftermarket metallic linings are available for both the early and the late models, as well as new Kevlar-based compounds. For the '65–'69 models, if the parts stores can't supply Corvair linings but do have Chevelle linings, you merely use front Chevelle shoes on the rear and the rear shoes on the front. Modify the front Chevelle shoes to fit the rear by adding the parking-brake holes and the notches for the cross-link and emergency-brake pickup. Make the Chevelle fronts look like your stock rears. Remove the front cylinders and bore these to 1.00-inch diameter for use with pistons, cups, and other parts from the '63–'64 Corvair metallic-lining wheel-cylinder-repair kits, No. 5461649, or an aftermarket equal. A special metallic brake master (main) cylinder No. 5464275 can be installed, but is not essential. You may also need minor items such as adjustment and pull-back springs, and brake-shoe hold-down pins, etc. All the metallic lining manufacturers recommended changing self-adjusting brakes to non-self adjusting setups with the metallic linings. Kits are available to do that changeover. As mentioned before, metallic linings have better friction when warmed up, self-adjusting brakes tend to adjust independently from each other, causing brake pulls as they start to warm up. The proportioning valve mentioned in the previous paragraph should also be installed. Drums should be new, carefully refinished or smoothly worn for proper functioning with metallics. GM intermediate sized cars of the late '70s early '80s (Monte Carlo/Regal) used aluminum drums (with steel friction surfaces) on the rear, to go with their front disc brakes. Those rear drums will fit the front of the late Corvair, but 14-inch diameter or larger wheels must be used.

Vented brake-backing plate on Thomas racer is cut out at front and rear to induce airflow over metallic brakes. Scoops at front would be a help.

Disk Brake Conversions (Choice #2)

Disc brakes conversion kits (front wheels only) were available in the 1960s for the 1960–64 models. Even then, their "modest" stock braking ability was noticed. But low demand caused discontinuance on the part of Cragar and others. In the 1990s new kits came on the market for the fronts of early models, and are still available. Late models have had various brake kits available for years, due to the road racing of the cars in various guises. The 5-bolt wheel pattern of the late Corvair makes it common with hundreds of other, larger, heavier cars. You can buy both front and rear disc brake kits for late models for street use. They even incorporate the emergency brake on the rear calipers, needed for street use. In general, street-type disc brakes weigh more than drums. If you are after that last bit of weight removal for autocross, stay with drums.

4-Wheel Disc Brakes for Racing

In the last 10 years, several companies have brought out kits to convert the GM muscle cars of the '60s and '70s to performance discs. Most of them use discs that are perfect for the Corvair. The floating calipers from the "Metric" era of GM are a good fit. Building a system means assembling things from different sources. This is the system I have on my car.

Front brake uses GM metric-design calipers on a 10.5" S-10 4WD truck rotor. Aftermarket replacement caliper is from Stainless Steel Brake Corp. It is a bolt-on replacement for the GM caliper.

The caliper is mounted behind the spindle on brackets from Scarebird Classic Brakes.

- Front Disc: Baer DecelaRotor 10.5" Diameter P/N 5550-020—Fits '97 Chevy S10 4WD-front
- Rear Disc: Baer DecelaRotor 10.5" Diameter P/N 5550-020—Fits '97 Chevy S10 4WD-front
- Front Caliper: Stainless Steel Brake Corp Aluminum Force 10 Sport Twin L & R–Twin 38mm pistons
- Rear Caliper: Stainless Steel Brake Corp Aluminum Force 10 Sport Twin L & R–Twin 38mm pistons

Notes:

- Front Caliper: brackets from Scarebird in Washington State.
- Rear Caliper: brackets from 1980-85 Cadillac Seville rear. L & R—slight redrilling for Corvair pattern.
- Same rotor spec all around, same caliper spec front/rear L & R.
- Master Cylinders: pair of ¾" Wilwood masters in parallel. 2 psi residual pressure valves.

- Rear hub assemblies must be pressed apart and back together.
- Front hubs, rear stub axles, outside diameter turned down to allow rotor to slide over.
- Centering shoulder for original brake drum removed F & R.
- ARP wheel studs P/N 100-7701 (set of 5).
- No emergency brake for now.

Regardless of the type of brakes you use, replace the standard fluid regularly if the car is in normal use. There is a constant debate over the merits of silicone brake fluids. If a car will sit for most of its life, and only lightly driven, silicone brake fluid, DOT 5, is a godsend, because it rejects water and won't damage paint. It can result in a spongy pedal feel. Polyglycol-based fluids, such as standard DOT 3 or DOT 4 fluids, have better compressibility when hot, so most racers run that. Recently a new fluid has come onto the market DOT 5.1—it is not silicone based, it is glycol based. Like DOT 3 and 4, the new 5.1 fluid can be mixed with other glycol fluids. The new 5.1 fluid has a higher boiling point than the DOT 3 or 4. It is more expensive however, and still absorbs moisture from the atmosphere. You can change the fluid to a heavy duty replacement by flushing the complete system with new clean DOT fluid of the same type. In a pinch the Motorcraft HD DOT 3 fluid is the best commonly available DOT 3 fluid and is available at any Ford dealer.

Dual Master Conversions

Kits are now available to replace the single master cylinder on the 1962 through '66 Corvair with a dual, inline master, similar to those used on the '67–'69 models, and most other cars since. The advantage of the dual master is retention of some braking ability, even if a brake hose or metal line is severed. This is especially important for the early models with swing axle rear suspension. If a U-joint/wheel bearing failure causes a rear wheel/tire/axle assembly to move outward from the differential,

This dual master cylinder installation is the stock '67–'69 install. The complete system, with a couple of minor fitting changes, can be installed on any 1962–66 Corvair. The Corvair vendors also have complete changeover kits available.

There is a new dual master cylinder on the market for casual racing applications. It places the front and rear masters next to each other and allows the front-rear balance to be adjusted from the driver's seat. This is similar in practice to the racing dual pedal assemblies, but bolts in place of the Chevy single master, onto the Corvair firewall. The unit is made by CNC industries of San Diego, California. Front and rear master cylinders can be sized for the brake calipers used, then outlet pressures balanced to prevent lockup at one end.

the stock brake system, including the emergency brake system, will be virtually worthless, because that drum will no longer restrict the shoe movement. A dual-master system would retain front brake operation in that situation.

Whether you use stock or replacement metallic linings, leave the outside of the drums in their rusty-iron finish. Or, give them a coat of non-insulating flat-black paint. Drilling the drums does not aid cooling, and polishing, chroming, or painting them white or aluminum is absolutely forbidden. Any of these changes will reduce the cooling efficiency of the drums. Incidentally, '64 Corvair rear drums have 40 radial cooling fins on their periphery. Vented backing plates are of little help in brake cooling unless scoops in the air stream are hose-connected to the backing plates so that air is forced through the brake assembly.

Wheels

Wheels present both a problem and an opportunity to the Corvair owner who wants maximum cornering performance for driving pleasure or for competition. The widest factory 13" steel-rim wheels available are those Chevy supplies on the car when you buy it, 5.5-inch. Increasing this rim width is essential to provide improved cornering power. Because this requires special machine work on the wheels, many owners never bother with the task, and they miss out on one of the finest improvements they can make to their chassis setup. Wider wheels can be constructed by using parts from two wheels to make one wider wheel. The inner portion and center of one wheel is welded to the outer rim section of another wheel. Or, one wheel can be cut in the center and a steel band welded in place to accomplish the desired widening. Because any width steel aftermarket wheels have become available, few people still widen the original wheels. There are many aftermarket wheels available for both the early and the late Corvair because of the commonality of the bolt patterns. Early models can use larger diameter wheels but are generally limited to six or six and one-half inch wide wheels. Many Datsun 240Z wheels were offered in the 14-inch diameter by 6-inch width. With the correct backspace and lug nuts, (don't use the Datsun lug nuts, they're metric threaded) these wheels will work fine on an early model. If you always liked the look of the Chevy "Rally wheels" of the '60s / '70s, they are easy to find for the 5-bolt late models. There is a company in California that will make a 14 × 6 Rally wheel in a 4-bolt pattern, with the right offset for the Corvair. With an outer trim ring and a pop-in center, it is a real '60s factory look. When changing wheel diameters, keep the overall tire diameter in mind. There are many aftermarket 13-inch wheels that will also work fine on an early model. The popularity of the early Japanese imports that shared the bolt-pattern has assured a good supply. The problem is poor availability of tires.

A quick discussion of backspace is needed. Backspace is the distance between the mounting face of the wheel (the point that bolts up to the brake drum), and the rearmost point of the wheel outer. When looking at an empty wheel, one with no tire mounted, you can measure backspace with a yardstick (or other straightedge) and a ruler. Set the wheel face down on the floor. Place the yardstick across the rear of the wheel rim, close to the largest diameter. Then slide the ruler down until the ruler end hits the mounting face. The amount shown on the ruler where it hits the yardstick is the backspace. If you perform this task on a wheel that has a tire mounted, your straightedge must sit down on the wheel's outer most points, not on the tire. All stock Corvair cars, early or late, have the same backspace—about 4.5 inches. Remember that the Corvair wheel width measurement (5.5 inches) is between the tire location points on the insides of the wheel flanges. The actual overall wheel width is about ½ inch to ¾ inch more. If you try to calculate "wheel offset" you will have a tough time in wheel discussions because the terms "positive offset" and "negative offset" are not universally agreed upon. A word about "wheel offset." In theory "offset" is measured from the inside of the wheel-bolt plate to the wheel-rim centerline. It is always a real number. The term "positive" or "negative" offset depends on the opinion of the person measuring. There is no agreement in the wheel industry or the automotive industry. Chevy's parts books mention offsets but neglect mentioning whether the offset is positive or negative, nor do they define it. The more acceptable wheel measurement is "backspace." When using 13-inch wheels on the front of the Corvair, backspace must be kept near stock to provide sufficient steering and ball joint clearance. Any additional width must be added to the outside edge only. Using 14-inch and 15-inch wheels provides more clearance and a higher backspace wheel can be used.

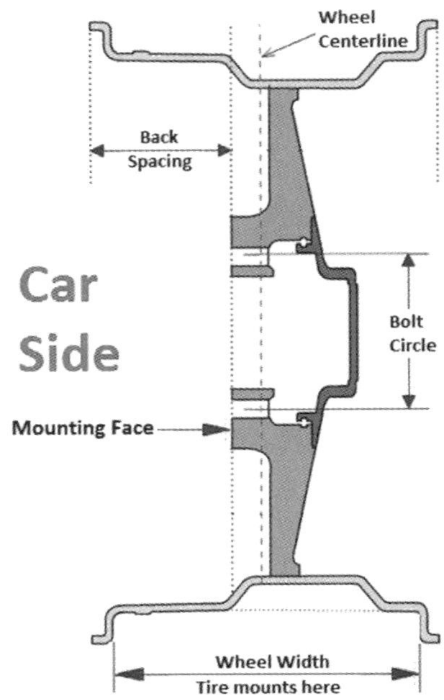

The serious decrease in the number of cars using 13-inch wheels since the 1970s, and the lack of a solid supply of performance street tires, has pushed many Corvair owners into larger diameter wheels and tires. Remember, your car doesn't care what diameter your wheels are. The overall tire diameter, however, is very important. Changes to the OD will affect speedometer calibration and brake loading.

Wheels up to 6.5 inches wide can be run on the '60–'64 model cars without clearance difficulties. Up to 8-inch wheels can be used on the rear of '60–'64 with some outboard fender flange flattening and crossmember denting on the inside. This is somewhat dependent on tire size and shape. Some grinding may also be needed on the inside to provide the needed clearance. '65–'69 models have loads of room under the rear fenders, so wheels up to 8-inch width can be easily accommodated. On the front, you may have to remove any extra threads of the front strut rod where these protrude through the control arm.

On wheels for competition, you can weld in a ⅛-inch-thick flat plate behind the spider, welding it at several points as shown in the photo—not

Chevy rally-style wheels have been produced in every size and bolt pattern. These 14 × 6 Rallys in a 4-bolt pattern were produced by Pete Paulsen's House of Wheels in California.

A seven-inch wheel width is plenty for most street applications and you can mount easily available, reasonably priced tires. With a 4-inch or greater backspace, they will clear everything. The TSW wheel shown is a 15 × 7. It is representative of dozens of available wheel styles.

The generous wheel wells of the late model body and the GM-common 5-bolt pattern, (5 on 4.75"), means easy wheel availability for larger wheels. Shown is a 15 × 8 aluminum wheel from TR Motorsports. With a 5-inch backspace, it clears everything, both front and rear, and the wide wheel assures an adequate supply of traction.

The 15 × 7 Camaro aluminum rim is heavy but rugged. They are hub-centric for tight mounting on the late-model Corvair. They were standard on Camaros through the 1980s. Many have outlasted the cars and appear at swap meets and the mini-ads all around the country. For the Corvair's 7/16" wheel studs, the correct "bulge nut" style lug nuts should be used. Stock Corvair lug nuts will cut into the aluminum giving false torque readings.

Don Eichstadt's 7-inch wide competition wheels for his '65 racer show extra band welded into front edge of wheel to add width (arrows point to added material).

Back side of same wheel with stiffener welded at eight points inside and around center of spider.

a continuous bead around the rim. When the plate is welded in, it will be necessary to knock, turn, or grind off the drum balance weights. Rebalancing may not be needed. Stiffeners to the outside of the rim can also be added, especially on 7- and 8-inch widths. Any wheels should be inspected prior to each competition event, regardless of the care that has gone into their construction. '60–'61 Corvair owners should buy a set of later wheels for racing or even for vigorous driving. The early '60–'61 wheels are very prone to crack around the bolt holes.

1965 wheels are better suited for competition with their larger 5-bolt mounting circle which spreads the loads more widely to reduce bolt-circle cracking.

Chassis tuning with wheel width is an accomplished fact. Many racers prefer to use stock-width or slightly wider wheels at the front, with very-wide wheels at the back. Thus some differential is accomplished, with the wide wheels tending to eliminate the oversteering tendency. There is plenty of room for experimentation to get the chassis perfectly mated to the type of racing you are doing.

Reversing wheel rims so that the major portion of the wheel is placed toward the outer edges of the car may look very stylish or "Kustom Kar," but the technique is not recommended for Corvairs. Fender wells will be fouled by the tires, and the rear axles on '60–'64s will be subjected to excessive loads which may lead to premature axle breaking. In addition, reversed wheels are not really needed because Corvair is already a "wide-track" vehicle, equaling or surpassing all recent Pontiacs bearing that designation. The '60–'64 Corvair tread is 50% of the wheelbase, '65s are 52%. Money that you might spend on reversing or chroming wheels can be better invested in at least two special wide rim wheels for the rear. Add them all around if your budget permits. Both wheel patterns, the early model's 4 bolts on a 4.5-inch circle and the late model's 5 on 4.75-inch circle are common industry bolt patterns. The 5-bolt pattern was GM only, but almost universally used (short of the bigger Olds/Pontiac/Cadillac models). The model year 2013 C6 Corvette still used the same bolt pattern for its 19-inch diameter wheels. A last note about bolt patterns. The GM/Chevy bolt pattern is equal to 120.65mm. Larger BMWs have used a 5 on 120mm bolt pattern for many years, with some years having similar offset wheels. People have just bolted the BMW wheels on with, seemingly, no ill effects. I don't recommend it, I have had several sets of "BMW-destined" aftermarket wheels re-drilled to match my 5 on 120.65mm pattern. The engineer in me won't allow the twisting effect on the wheel studs. Speaking of wheel studs, if you need to run some wheels that need spacers to help on clearance, you may need longer studs. ARP makes some nice ones that have a long unthreaded lead to help the lug nuts go on smoothly.

Because '60–'64 axles have a tendency to break when given the rigorous use of competition, you may want to install the stronger axles used on the Greenbriar. Installation of the longer axle requires adding a spacer between the control arm and the Greenbriar bearing (larger than passenger car). The 5-bolt flange extends farther out than the stock axle, so you may have to relocate the wheel spider in the rim to maintain fender-well clearance on very-wide rims. Part No. for the axle is 3771921. The mating bearing is No. 7451231. The parts are not cheap, so check in a junkyard to see if you can find them there. Incidentally, all axles, whether new or used, should be regularly checked with Magnaflux or other crack-detecting solutions, even though this means pulling the bearing off to check the area where the flange joins the axle. 1965–69 axles have not given any problems in severe racing usage.

The specialty wheel manufacturers never "got with it" to make wheels for Corvairs. Aluminum or magnesium wheels with only an extra half inch or inch of rim width are not a sensible purchase when you can buy 7-inch-wide aluminum or steel wheels, or have wider steel wheels made up specially. Don't rely on the speed shop's catalog jockey or online listing—measure the wheels yourself before you buy. Although there are wide aluminum/magnesium/steel wheels for the 5-bolt '65–'69 Corvair pattern, keep the 4.5-inch backspace (minimum) configuration in mind to avoid overloading wheel bearings and to keep the outer edge of the wide tire tread from being ground off by the fender-well edge. Only a few of the aluminum/magnesium/steel wheels have the 4.5-inch backspace. Most are lower backspace and you cannot successfully use "deep dish" wheels on the Corvair. Chevrolet Camaro 7 × 14 steel wheels, like the "Rally" wheel style (about a 4-inch backspace, 5-bolt pattern) will probably work in most instances with a 7-inch tread in front or 8-inch in back, provided that a 24-inch diameter is not exceeded in front or 26 inches at the rear.

Besides the obvious visual differences, the installation of larger tires and/or wheels has other effects.
A larger tire and wheel will usually add more weight. This will affect braking and has other gyroscopic effects. The late Corvair wheel well will accept a 7-inch wide by 15-inch diameter wheel without a problem, providing the backspace is close to stock. The 1982–92 Camaro 15 × 7 alloy wheel clears pretty well on the late Corvair, looks great, and there are thousands available at the nation's swap meets. The reason that wider wheels can cause problems is because of the inner clearance issues around the rim. The steering arms and tie rods, which ride inside the wheel area, are an obstacle which must avoided, so most of the additional wheel width ends up on the outside of the wheel. This changes the offset of the wheel, but, more importantly, it changes the steering geometry of the front end. If a slightly wider wheel with exactly the same offset as stock is used, steering geometry remains close to stock. But, as the centerline of the wheel moves outward, relative to the steering axis, reaction to suspension movement changes. Instead of a bump being handled by the upward movement of the suspension, to be absorbed by the spring and damped by the shock absorber, the wheel will try to turn outward in response to the bump. Unless you hit an identical bump with both front wheels at the same time, the car will pull in the direction of that affected wheel. This is the "nervous" effect of large tires and wheels. On the racetrack, the adrenaline rush will prepare you to deal with those effects. But on the street, you will run out of adrenaline and just be tired of fighting the wheel. This is the reason behind my suggested seven-inch wheel width limit for street use.

Tires for Corvair

It should not surprise you to learn that stock Corvair tires were not the world's greatest for vigorous driving or closed course competition. Their construction would not allow safe use of the car's full performance. The knowledgeable owner replaces his stock rubber with race-bred tires at the first opportunity, mounting these on wider rims for dramatically improved road and competition handling characteristics.

The factory-supplied two-ply tire might have been ideal for Mr. Average Driver. Strict factory acceptance tests proved that the tires were excellent in respect to economy of operation, rolling resistance, and car stability. But at the factory-recommended inflation pressures of 15 front and 26 psi rear, the tires will roll under the rim when cornered hard and provide less than desirable stability when the wind is blowing. Of course nobody is doing any kind of enthusiastic driving on those stock-type 2-ply tires today. Even so, any tire can be improved for high-speed driving of all kinds, including slaloms and autocrosses, by merely increasing the inflation pressures from 5 to 10 psi over the factory specification. Increasing inflation pressures stiffens the tire sidewalls against rolling under during cornering, but, as you probably guessed, this is only a compromise measure. Pumping up a tire makes it rounder in both dimensions, reducing the rubber-to-ground contact patch that is available for traction.
So, while pumping up the tires can increase car stability in cornering and its resistance to wind-wander there is less rubber available for stopping or accelerating. Increasing tire pressure beyond the point where the sidewalls are stiffened reduces cornering ability because then wheel hop begins. Stiffening the sidewalls can also be aided by installing tubes into the tubeless tires, but is no longer recommended by any tire manufacturer. In finishing our discussion of original-type tires it should be mentioned that the best traction on dry surfaces will be obtained with tires which are nearly worn out. If only a tiny fraction of the tread imprint is showing, then there will be the maximum amount of rubber on the ground for tractive and cornering effort.

Tires have changed considerably since the 1960s, mostly for the better. Original size tires (including period-style whitewalls and correct markings) are available from the restoration industry. If your stock class show car needs tires, Coker tire still makes a 6.50-13 tire that is quite close to the original. If your car originally had the 7.00-13s, you are out of luck for now.

But, unless you are intending to have a "show-only" car, you should install newer, radial tires. Exact-size replacement radial tires are getting harder to find because newer cars have been moving away from 13-inch tires, even on the smaller-size cars. The early model cars can use a 185/70-R13 size for most models. The later cars can use a 195/70-13 size, or a 185/80-R13. The Corvair vendors have stepped up buying large enough quantities to keep the tires in production, MAXXIS is still producing the 185/80-R13 in an appropriately skinny whitewall. Although slightly smaller than stock in diameter, these are close approxima¬tions to stock tires and will fit fine on original wheels. Many people feel that the Corvair, especially the late model, "looks" under-tired. They feel that a stock tire doesn't fill up the wheel well. Because many 14-, 15-, and 16-inch wheels are available and will bolt on, it is a common modification for late models to install taller wheels and tires. The wheel well on the early model Corvair doesn't allow tires or wheels too much wider than stock. If you are making the jump to larger diameter tires, a new world of tire availability opens up. Within the limits of wheel width, overall diameter and wheel well volume, wheels as large as 17 inches diameter and tires as wide as 8 inches can be fitted onto the Corvair. Late models can run 225/50-16s all around,

Lined up and ready to go, these late models are running the real race tires necessary to win in the Prepared Classes. Hoosier Racing Tire Corp. and Goodyear both produce "slicks" in the USA for these race cars.

as much tire as you will need for any reasonable performance driving conditions. Even taller wheels – (with even skinnier tires) can be fitted to the late Corvair for "cosmetic" appeal. Be careful of wheel well clearance.

Competition Tires

Back in 1965, Bill Fisher recommended real racing tires of that era as the hot setup for the street. That is no longer an option, nor is it needed for the street. There is an excellent selection of performance tires for our cars. About the only item in short supply is a larger 13-inch tire. A 205/70-13 was available and still might be found if you look hard enough. They are not generally available, though. The more commonly found 205/60-13 size is quite short for Corvair use. Unless you are running 3.27 rear gears, highway travel will be noisy and fuel consuming. If you search them out, even shorter 13"tires are available.

In the last 15 years, several tire companies have introduced new tires that are intended for "competitive" events, but still meet DOT requirements. Each year the market becomes more competitive, and tire companies introduce new tire lines and discontinue old ones. BF Goodrich introduced several lines of tires, Yokohama, Hankook, Falken, Goodyear, Hoosier, and Kuhmo tire companies have all introduced special sticky (and DOT legal) tires to meet the demand for better grip under optimum conditions. The Kumhos, Falkens, Yokohamas, and BFGs can all be run as street tires, as the underlying carcass of the tires approximate their regular tire construction, with sticky-soft rubber compounds in the tread. The Hoosier is the stickiest at the autocross track, but quite fragile for anything like street use, and they have almost no tread. I don't recommend them for the street. The best street combination I have seen in the last few years is a 205/55-15. Several different manufacturers make this size, they fit fine on a 7" or 8" wide wheel, and they are the right height to match the gear ratios on the Corvair. An extra bonus, their height matches the 225/50-15, if you want wider tires on the back.

Race Tires

Real race tires are still available for folks who want to race Corvairs. Many open-wheel race cars still run 13-inch-diameter wheels so both Goodyear and Hoosier have lightweight, super sticky tires waiting for you to write the check. There are plenty of 15-inch tires that will work on the Corvair as well, and most organizations allow swapping to larger diameter wheels and tires. In the 1960s, Goodyear used to subsidize race tire purchases with street tire sales. They no longer do that, and Hoosier has only a tiny street tire market. So race tires are expensive, although there is a complete aftermarket in used race tires because the deep-pocket racers often replace them prematurely. They must feel it is the tires making them slower, as it couldn't possibly be the driver. Do not be tempted to run real race tires on a street car. Not only is it illegal, it would put you in a dangerous position in rainy or other inclement weather conditions.

Recently, the Sports Car Club of America standardized performance tires with a 200 wear rating (minimum) as a requirement for their street classes for Autocross competition. Several other sanctioning bodies have also adopted that wear rating. Most of the performance tire manufacturers have brought out new tire models to compete in those classes. They are available in some very wide 15- and 16-inch sizes. The 200 rated tires can be street driven as well. They will wear more quickly than todays 400 to 500 rating mileage specials, but they will be more fun to drive. The Uniform Tire Quality Grading system, including this wear rating, is self-regulated (the tire companies test their own tires and rate them). There has been some question as to how accurate the mileage ratings are on these maximum performance tires. Long life and long mileage is not their reason for existence. If you intend to buy these tires for competitive events, search around for performance comparison tests, both in magazines and online. The manufacturers are constantly bringing out new models, trying to one-up their competitors. Two great sources for this information are Grassroots Motorsports magazine and Tire Rack, a major online tire retailer.

Tire Pressures

The only lawsuit ever lost about Corvair handling was because of incorrect tire pressures having been set by a dealership employee. Whether or not GM should have used a design that was so dependent on correct pressures to preserve handling is beyond this debate. What is not beyond debate is the fact that to preserve the predictable handling of the Corvair, you should maintain the tire pressure differential between front and rear tires. Why? The Corvair weight distribution is different from almost all other cars, certainly all other American cars. With a front-engine car the majority of weight is on the front, steering, pair of tires. On the Corvair, the majority of the weight, usually around 60%, is over the drive wheels, in back. This is one of the factors that helps the Corvair handle so well in poor weather conditions such as snow. With the recommended amount of air in the stock front tires, around 15-18 psi, the full tire contact patch is in touch with the road, and the steering input results in tire traction. If the tire is pumped up to 32 psi, common for new cars today, very little of that tire patch will be touching the ground, and the car will not respond well to steering input. Radial tires, due to their softer side walls, have limited this problem, somewhat. But I still

Before going out on course at an autocross, you can mark the outside edge of your tires. Here I have used white shoe polish. After your run, the polish will be removed where the tire has made contact. You can adjust pressures depending on the amount of contact.

maintain a 10–12 psi pressure difference between the fronts and rears, with the rears having the higher pressures. The pressure decal, originally installed on the back face of the glove compartment door, at least on late models, is a reminder of this differential. The suggested pressures quoted are for the original non-radial tires. The new radial tires have become industry standards over the past 30 years and they work fine on Corvairs, providing they are sized appropriately for the weight of the car or truck.

No matter which tire you choose, you will have to experiment with pressures to optimize their performance. Remember to keep the rear tires about 8 to 10 psi harder than the fronts. Goodyear suggests maximum pressures of 32 psi front and 40 psi rear as starting points for their race tires. They have found that the fastest lap times are recorded with the lowest-possible inflation pressures that will maintain vehicle stability. This is because the lower inflation pressures increase the size of the rubber patch that contacts the ground. Race tires excel in their ability to provide maximum stability at low pressures, due in part to the fact that these tires have a tread about one-inch wider than most other tires of equivalent section. So, experiment by dropping the inflation pressure from their starting recommendation. Real tire technicians use tread temperatures to "tune" the tires. They measure the tread temps on race tires as they come off the track. The tire is working best when the temperatures are the same across the face of the tire, and within their design range. You can vary suspension settings and pressures to achieve that. For those of you without a tire pyrometer, white shoe polish or even chalk can point you in the right direction. Before you go out for your first run at a slalom, use the chalk or polish to place 4 or 5 short stripes aimed at the center of the tire but going over the shoulder. The white marks should cover at least from the sidewall to a couple of inches of the tread. After your first run is complete, you can examine the chalk to determine how hard the tire worked. The chalk should be scrubbed away all the way to the edge of the tread, almost to the sidewall. If you are not using all of the tread, let some air out and let the tire roll over more. If the chalk is gone on the sidewall, time to add air!

19

Corvair Engines in Dune Buggies, Porsches, and Volkswagens

Dune Buggies and Kit Cars

Most dune-buggy builders of the sixties and seventies used a shortened VW chassis with the VW front and rear suspension and a VW engine and transaxle. While a stock or modified VW engine was still the most popular, the prices for these engines and transaxles have zoomed on the used market, causing many buggy enthusiasts to consider Corvair power. Corvair engines, especially older ones, will probably continue at more reasonable prices because they must be adapted to fit the VW transaxle. The idea of adapting does not appeal to Mr. Average Man, regardless of how well engineered or how many such adaptations have been made to lead the way. However, if you are about to build a dune buggy or bought an old one in the Mini-ads and are still trying to decide between a Corvair and a VW, let's look at the differences to help you make up your mind.

Nine Corvair engine models range from the 140-cubic-inch 80 HP 1960 model through the 164-cubic-inch 95, 110, and 140 HP models which were offered in the 1969 models—and the '65 and '66 Corvair Corsas had turbo-supercharged 180 HP units (150 HP in the 145-cubic-inch 1963 and 164-cubic-inch 1964 Spyders). None of these engines are especially better than the others except that the 1964 and later models are much larger in displacement than most VWs. Aftermarket VW engines are now available, in sizes close to that of the Corvair, but at a much higher price.

This Meyers Manx has a breathed-on 140 HP engine.

Rebuilding costs for the two engines are similar, but the VW requires several "trick" modifications to get reliability for competition or off-road use. Hot rodding the VW to obtain 100 to 110 HP (comparing advertised SAE HP ratings) costs thousands, according to expert VW tuners. You can get this much or more from a stock Corvair without spending a penny for modifications. In comparing costs for the two engines, you ought to consider that the Corvair requires the Crown adapter parts, and if you will be using a '68 or later non-swing-axle VW or Porsche transaxle, the costs of a reverse-rotation camshaft must also be included in the Corvair column. And if you are starting with a 1963 or earlier Corvair, the reverse-rotation engine requires the straight-bladed blower fan. Weight difference between the two engines is about 75 pounds, and with the big tires that you will undoubtedly be running on the rear of your buggy that extra weight will not even be noticeable when comparing handling. This brings up a final point—the VW engine, even when modified, requires new 3rd and 4th gears to make these cogs usable with big tires. Add to this the fact that anything you do to the VW engine to raise its horsepower (except for the expensive process of replacement

Bruce Meyers and Crown Manufacturing teamed up to build winning cars.

Photo from *Hot Rod Magazine* illustrates use of VW spur gears at outboard ends of Corvair half-shafts. Corvair engine is located amidships and a Powerglide transmission is utilized.

big-bore cylinders and a stroker crank) will narrow its usable RPM range, moving it upward on the scale so that low-RPM performance falls way off. By the time you have purchased one or two hop-up items for the VW you could have purchased a Corvair with twice the horsepower and with the adapters, clutch and flywheel, starter adapter, and even a reverse-rotation camshaft. Thus, it is hard to argue for the VW. Of course, because VW is no longer making their air-cooled motors, used VW motors are getting even harder to find than the Corvairs. There are several suppliers of "built from scratch" VW motors, but those motors are pricey.

Seasoned off-road racers remove the cooling-air damper ducts from the Corvair's underside, thereby decreasing oil temperature about 20° F and reducing vulnerability of the engine to rock damage. Of course, this is not recommended for a street buggy for reasons detailed earlier. The distributor is reasonably well sealed against dust so long as the vacuum-advance boot and point/cap dust shield are intact. Air intake should be through a flexible hose attached to an air cleaner just under the roll bar. The Tri-Phase construction-equipment-type air cleaner, which spins the entering air to drop out dust, seems to work best of all. If K&N type pleated filters are used, these must be oiled. Carburetors need throttle-shaft seals. Several manufacturers have inexpensive kits with O-rings, washers, and springs

Support straps around front and rear of transaxle control excessive movement found in stock mounts, preventing transaxle breakage in off-road use. These were Crown parts.

to seal four carbs. O-rings are placed at each side of the carb base on the throttle shaft. A light spring between the lever and a flat washer holds the O-ring in contact with the carburetor for an effective seal. Verify that the throttle returns to idle correctly.

Late-model VW transaxles beginning in '68 (with the buses and semi-automatics) with two U-joints per axle cannot be used in reverse rotation because the internal parts are designed for only one direction of rotation of the input shaft. Some stick-auto transaxles with the "tunnel-type" case can be reversed by flopping the ring gear. These are identifiable by the same-size bolt circles on each side of the differential, indicating that the ring gear can be installed from either side. Non-symmetrical cases with only one large bolt circle cannot be reversed with a flopped ring gear. If the chassis dictates it, or the transaxle you want to use cannot be flipped, you will have to reverse rotate the Corvair motor. Reversing the motor requires almost a complete disassembly, but it is not tough technically. Many Corvair engines have been reversed. A few items will have to be replaced. The camshaft is different, the distributor/oil pump drive gear that mounts on the crankshaft is different as well as the companion gear that mounts on the distributor main shaft. The gears are swapped to a reverse-cut pair because the distributor and oil pump must turn in their original directions to pump oil through the motor. After reassembly, a few things will be different. To run correctly, the ignition wires must be reversed in the distributor cap. The new firing order will be 1-6-3-2-5-4. The fan belt, and the fan, will be traveling the opposite direction. Check the guides and the pulleys for alignment. You should change out the alternator fan/pulley to the regular Chevy style, because the Corvair fan, designed for operation in the other direction, won't pass enough air when spun backwards. A reversed starter should be part of the adapter package, along with the modified flywheel and cast adapter.

Corvair-engined buggies need rear motor mounts. These can be suspended from an encircling tube or bumper around the back of the buggy. Such a structure can be tied to the chassis and roll bar and can support both the rear motor mount and the skid plate. If a

Triangulated roll cage with hoop around back of engine provides good rear mounting for the Corvair powerplant in this buggy by Barbary Coast Enterprises of South San Francisco, California.

Skids extending under this Corvair protect it from rocks and provide a rear support. Both are important factors when using the Corvair engine in a buggy.

Funco's all tubular steel chassis is ideal for use with Corvair power because the rear engine mount can be hung from rear cross tube. Additional tubes under the engine can mount a protective skid plate as a rock guard. VW suspension is used at front and rear. These chassis have been extremely successful in off-road racing.

VW transaxle is used, then extra trans. mounts are also needed. Mid-engined buggies are easier to equip with engine transaxle mountings, and with skid-plate protection for the engine.

Corvair Suspension for Buggies

Corvair suspensions are now being used in some dune-buggy chassis. The very high rate of the stock Corvair springs (designed for the weight of a complete Corvair) is exactly the opposite of the suspension requirements for off-the-road fun and competition. A softly sprung, long-travel suspension is needed, and this cannot be obtained with the stock Corvair springs. Although the soft rear springs offered by Crown Manufacturing for its Corv-8 conversions can be adapted to buggies equipped with '65 and later Corvair rear suspensions, even these springs are too stiff for best off-road use. There are now commercially available springs soft enough for using the Corvair front suspensions for off-road use. You can choose free length, spring rate, and end type. You can use the Corvair suspension fore and aft for a street buggy, but only with the specially made springs. For serious off-road use, springs of about half the regular spring rate would be required. One dune-buggy manufacturer that used Corvair components modifies Chevelle springs by cutting off two coils, but admitted that they were way too stiff for off-road use. It should also be noted that the Corvair spindles make a much stronger front end than can be obtained with the original VW components.

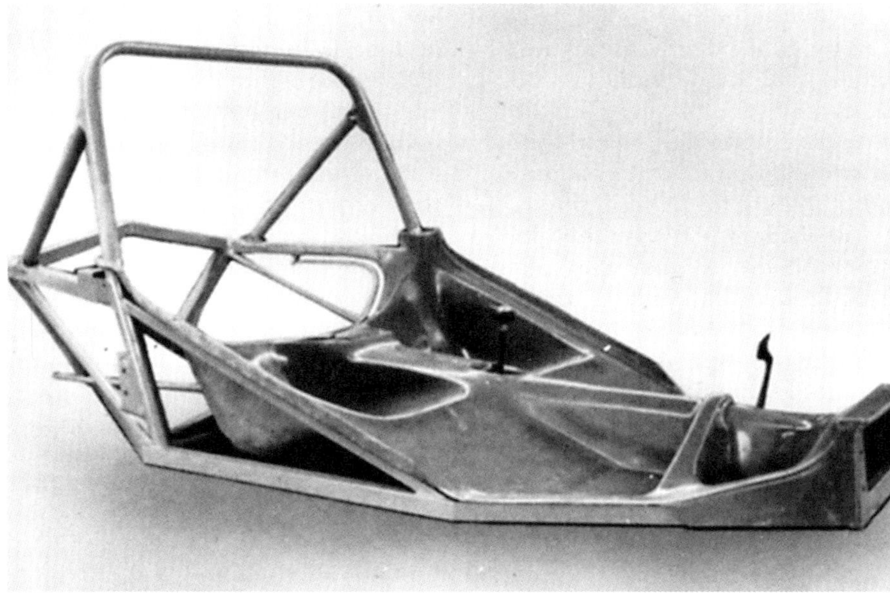

Deserter GS tubular space frame with shift linkage and fiberglass seat pan/floor installed. Roll bar is an integral part of the chassis. Suspension is stock VW at front, coil spring over shock at rear with swing axles. Brakes can be Porsche, or VW drum, or disc.

Mid-engined Deserter GS with engine cover removed to show Corvair 140 HP engine with Holley four barrel. Engine is Crown-adapted to VW transaxle. Engine cover attaches with six Dzus fasteners.

Mid-Engine Buggies

The rearward position of the engine mounted behind the transaxle in the usual dune buggy poses problems in serious off-road activity: the engine can hang up, perhaps leaving the rear wheels off of the ground—and more likely, causing severe damage to the engine and/or transaxle. While the rear mounting produces maximum cornering power with the right tires on shorter slalom courses (but makes it harder to drive), and the engine location aids in traction for some off-road situations, a mid-chassis location for the engine offers some advantages. First, the weight is better distributed, especially for high-speed events. There is less polar movement on long, fast courses—and less understeer on short courses. Secondly, the overhang is limited to that of the transaxle—less than half of that when the engine is in the rear. Thirdly, the engine is better protected from damage, gets less dust and is better protected from drowning out in water hazards, although airflow for cooling must be planned.

Deserter GS with engine cover installed.

Corvaired sand dragster front mounts '61 engine to a Chevy pickup gear box driving to Henry J rear end with 4.11 gears. Two Ford carbs are on fabricated manifold. Pete Sukalac Photo.

In 1969, several buggy manufacturers had decided to market mid-engined buggies capable of handling Corvair power. One of these, Dearborn Automobile Company of Marblehead, Massachusetts, offered the Deserter GS (GS stood for "Giant Slalom"), which is pictured and described in the photos and captions. Those pioneering firms were joined by an ever-increasing number of individuals who were "scratch-building" buggies from the ground up. If you are planning to buy or build a buggy, observe the trend in almost all auto racing, Detroit's idea cars, Crown's Corv-8, and dune buggies such as the Deserter GS—think seriously about traveling the mid-engine route.

The Corvair engine can be used in its stock form (no reversing is required) by merely mounting it to a built-up VW swing-axle transaxle and using Crown's engine-to-transmission adapter, clutch, flywheel, etc. The entire shifting mechanism (linkage, etc.) was available from Dearborn Automobile Company. Today you would have to fabricate it, or look for aftermarket Porsche parts.

Some readers will wonder why the Corvair transaxle, with its extra beef, cannot be modified to turn backward as can the VW swing-axle transaxles. The VW gears are straight cut and will operate in either direction, so the drive gear can be moved to the other side of the pinion. The Corvair hypoid gears are designed for one direction of rotation and cannot successfully operate in reverse rotation, although several V8 kits have used this feature. The Corvair transaxle can be used if VW Transporter spur-gear housings are mounted at the outer ends of the axles to reverse the "backward" output. This setup allows normal rotation of the engine and transaxle in a mid-engine buggy. Careful adaptation and sealing is required in grafting the VW closed tube axle parts to the open-tube configuration of the Corvair axles.

Porschevair and Volksvair, the Concept

In the early 1960s, Porsche 356 models and all Volkswagen bugs were inexpensive cars. The VWs of that era, with a few exceptions, have remained so. But the 356 Porsches have, like many other sports cars of that era, skyrocketed in price, as long as the cars have remained stock. Bill Fisher felt that his Super 90 roadster was so underpowered it needed an engine transplant. I cannot fault Bill for this decision. Today, anyone who completed that swap should have his financial head examined. It would severely devalue a pricey classic car. As noted above, aftermarket air-cooled VW motors, most of which would slide right into a 356, can put out as much HP as any stock Corvair, with no modification to the car needed. The original motor could be set aside for reinstallation if need be. That said, I believe that Bill's decision to swap out the motor, and his discovery of

Ted Trevor on his way to win under-3-litre class at 1966 Pikes Peak Hill Climb. Powered by 140 HP Corvair coupled to VW transaxle through Crown adapter parts which Trevor made, chassis is essentially VW under fiberglass dune-buggy body shell. His faster, Spyder-powered car, driven by Don Wilcox, lost a coil wire. Both cars were faster than most of the championship cars. Results for '67? No sportscar class was permitted!

the Corvair motor for that purpose, is what caused this book to be written at all. For that I am grateful. In deference to Bill's original design, I am including a modified/updated version of Bill Fisher's original coverage of the Corvair engine to VW/Porsche engine swap. After all, I did this swap, too. I swapped a Corvair engine into my Lola T320 SuperVee, replacing a broken Type IV VW motor. Probably a dumb financial move, but lots of fun!

Engine Swapping for Porsches, VWs, and Karmann Ghias

Want more "beans" for your Volkswagen or Porsche? Certainly you do. I know because I have owned both. Even the owners of the terribly expensive two-liter Carrera Porsches and early Porsche 911s will be forced to admit that their cars' performances hardly equal that of US-built machines costing half as much. General Motors has come to the rescue of every Porsche or VW owner who has had his stomach, heart, and wallet turned inside out by the parts and labor bill for major engine work. With the introduction of the Corvair, GM unwittingly provided power for those who wanted it. In 1962 I had decided that my own '61 Porsche Super 90 roadster was never going to provide the kind of excitement which I wanted in the acceleration department. So, I bought a set of the first Corvair/VW conversion parts which were made available on the West Coast. The fine parts which are now available from Crown Manufacturing Co. (Now Clark's), make those early parts a joke by comparison. In 1962–63 I purchased a '61 Corvair engine from a wrecker and proceeded to install it. Along the way I ran into all sorts of misinformation, together with traps, swindles, and pitfalls as related to the Corvair engine, accessories, and adapters. A lot of that experience has been set into this book. By the summer of 1964 when I was ready to build a hotter engine, Ted Trevor, owner of Crown Manufacturing in Costa Mesa, California, had really thought through the problems and was preparing to bring out the parts which would make these installations both simple and "sanitary." His parts fit and his instructions offer far more detail than you'll read in this chapter of the book.

Parts Suggestions

Get a 1965–67 engine, preferably the high-performance 140 HP with its big valves and four carburetors, or get a '65–'66 180 HP Turbo engine. The alternator used on '65 and later models offers a definite weight advantage over the engines with generators which were supplied through 1964. If you find a '64 engine available at a better price, it is a good second choice. The '64 95 and 110 HP engines are the same as the '65–'67 with the exception of wider head gaskets and the alternator. 1961–63 models give away 19 cubic inches of displacement which can only be regained by buying the crank, rods, pistons, and cam designed for the 164-inch engines ('64 and later). Be sure to get the starter and voltage regulator with the engine, the sheet metal shrouds which surround the engine, and the complete exhaust system.

Discard the flywheel and flywheel housing. If you are installing the engine into a VW sedan, bus, or Transporter, remove the oil-filter/generator adapter. Saw off the filter and plug the two exposed holes with ¼-inch NPT pipe plugs. Remove the relief valve from the adapter. If you want to use an oil filter, modify a '60–'61 adapter to permit installing ⅜ NPT to ½-inch hose fittings. Smaller hoses cannot be used as they would restrict the oil flow. The reason for using the '60–'61 adapter is that it is of heavier construction than the later models, which are die-cast. Porsches can use the filter in its stock position by cutting away part of the inner body shell at the back. If you plan to use a cable driven Porsche tachometer, read that paragraph before proceeding further. Remount the adapter and install the alternator or generator. Rotate the distributor to clear the deck lid on buses or Transporters. Move the plug wires to new positions in the cap to compensate and raise the coil two inches in its mounting bracket.

Heaters are important. When putting the engine in a Porsche, use Crown's heater-hose adapters, or make your own, to avoid making extensive modifications for hose outlets to clear the transmission mount tube. Retrieve the stock Porsche relay arms, one for each side, as these are needed to operate the heater valves which mount forward of the engine under the car. Remember to leave access for the bolt hidden in the left shroud. VW heater boxes with their control valves can be grafted onto the front of the stock shrouds if clearance is allowed for installing the engine into the car, These items are usually fitted during a trial installation of the engine to check fits, make sheet metal or neoprene shrouds, get the throttle linkage sorted out, etc. Seal the large hole on the front of the blower housing. This will ensure that the engine gets all of the cooling air it is supposed to have. Remove the exhaust crossover pipe. Refit this by cutting and lengthening or use the Crown part to save time on what is a really fussy fitting job, even though it looks easy at first. The crossover pipe must clear the underside of the transmission when installed.

The throttle linkage is set up with a relay arm on Porsches. The steel Porsche throttle-pull rod will only mate with their Part Number 616.08.013, so buy one, if you can. The VW throttle installation is accomplished by repositioning the lever on the Corvair crossover linkage rod. Move the lever to the left of the hump and angle it back—about 15 degrees. The lever is silver-soldered to the rod and can he moved after heating the solder until it melts.

Electrical–Easy Switch to 12 Volts

Because the Corvair starter functions on 12 volts, your conversion must include a complete changeover to 12 volts. 12-volt bulbs and flashers are available at any automotive supply house. The only special bulbs are the dash lights and dome lights which are available from VW or Mercedes dealers. A dropping resistor must be used for the wiper motor (4.5 Ohm, 10-watt) and radio (7.5 Ohm, 10-watt for transistor types).

Transaxle Modifications

If you have been wondering about transmission and rear axle life with the Corvair conversions—put your mind at ease. Off-road preparation can make the VW transaxle at least as strong as the original Corvair. So many of these conversions have been made that it is now common knowledge that the parts will last. Just remember to start in 2nd gear. Do not use full-throttle when starting in 1st gear, and do not spin the rear wheels. If you ease away from stops before applying full power, the parts will last much longer.

Corvair engines rotate in the opposite direction to Porsche or VW engines so the ring gear must be turned around. This is easily accomplished in the symmetrical differential housings used in earlier VWs and pre-1966 Porsches. Take the transaxle assembly out of the car, keeping close track of the number of shims which are removed at the front of the transmission as these determine rear-suspension toe-in. When you take the axle housings off, identify them. Keep the shims (gaskets) with the respective housings. No internal changes are required for Porsche transmissions in good working order. VW transmissions of the pre-1961 (or pre-1960 Transporters) with non-synchromesh low gears should be changed for the later style. If inspection

Top view of adapted Porsche transaxle with Crown Corvair starter adapter. Hose fitting leads to catch-can breather in engine compartment of author's car.

Careful installation of ring-gear carrier eliminates cracking indicated by arrow.

Close-up of adapted Porsche case showing capscrew for relocated deflector, NPT plug closing original hole, end of cut-off Corvair starter housing, and offset bushing for end of starter shaft.

indicates the need for new bearings, be sure to install them. If there is a differential breather, the pipe must be modified to clear the "flopped" ring gear. On my own Porsche, I heated the nut to melt the silver solder, then slipped the nut down on the breather tube, effectively raising the tube to clear the ring gear. Some Porsches have a deflector to protect the rear oil seal from being continually flooded with oil thrown off of the ring gear. Move this to the opposite side of the case so that it will continue to function as it did originally. Plug the original hole with a ⅛-inch NPT pipe plug. Drill a new hole to relocate the deflector. Use epoxy on the bolt to ensure against leakage.

The ring-gear carrier parts demand special attention. You will usually discover a crack at the location shown in the photo. This crack is caused by peening the spider-gear shaft pin into the casting (forging), thereby placing the part in tension and leading to cracking, When you reinstall yours, shot-peen the carrier parts for extra strength. Do not peen (with a hammer) the pin when you install a new pin to retain the spider shaft. Merely make a few pin-prick marks at each end so that the pin will not fall out, It is not under any load. Some VW ring gear carriers can be used in a Porsche if you use the latest parts. Compare the new carrier with your old one to be sure that there is clearance for the transmission shaft.

When you reassemble the ring gear into the case, get help from a Porsche or VW mechanic. This job is easiest on Porsches with the tunnel-type transmissions, 1958–65, and VW tunnel-type transmissions from '61–'67. Volksvair instructions give complete details for beefing up the VW transmissions and give helpful hints relating to reversing the ring gear.

Adapter Parts—Only Crown's Will Do

The adapter parts which you will need are the engine adapter, a flywheel with a ring gear which mates to the Corvair starter and permits use of the Porsche or VW clutch, a Corvair starter adapter with offset bushing, and a beefed-up clutch pressure plate

Two mechanical engineers put heads and resources together to build this Volksvair. To a rolled-over '55 VW sedan they added Porsche brakes and a junkyard 98 HP Corvair engine. Bluestreaks with McCoy caps on 5.5-inch rims at the front, 6-inch at the rear are used in conjunction with an EMPI bar at the rear and two stock anti-roll bars at the front for surprising agility on autocross courses. Since its introduction in March 1964, the car has been a formidable competitor, often defeating factory Cobras and modified Corvettes for overall top time of the day at autocross events.

Owners/drivers Glen Fotre and Erick Ressler of Santa Clara, California gathered the hardware shown here in the first seven months of competition.

Sole modifications to the engine include an Ansen 4-carb kit and headers.

During 1965–66 slalom seasons, Don Wilcox in the Crown Volksvair dune buggy won 25 out of 26 Top Times of the Day, running against 250-350 cars at most events.The field he competed against usually included 20 Cobras, 70 Stingrays, Lotus Super 7s, and many specially built slalom cars. Car had 140 HP Corvair engine, Crown-adapted VW transaxle. Basically was is a VW chassis with 65% of weight at rear. This is the same car driven by Ted Trevor to 1966 Pikes Peak win.

Bill Mece of Costa Mesa, California runs VW with stock 140 HP engine to get 98 mph and 13.8-second elapsed times at quarter-mile drags.

with disc. Although Crown used to supply adapted flywheels, they now furnish brand-new 6-lb. (that's right, six pounds!) flywheels which are aluminum castings mounting special ring gears, and with steel plates for the clutch disc surface. These have been proved in more than two years of all-out autocross and hillclimb racing in the Eric Ressler and Don Wilcox autocross specials. If you will be racing, a special 2,000-lb. modified VW clutch plate and disc should be installed. Install a new pilot bearing in the end of the crankshaft.

Install a new seal into the adapter from the counterbored side, packing the seal lips with high-temperature LubriPlate (available from Ford dealers). Remove the pan and the old pan gasket. Install the adapter with a new gasket and socket-head capscrews (supplied with kit). Loctite the capscrews so that they will not loosen. Liberal use of a gasket compound on the paper gasket is recommended as you are sealing the oil cavity of the engine, Before you install the pan, consider adding pushrod-tube baffles and/or a pan baffle as discussed in the "Lubrication, Cooling, and Breathing" chapter. If the car is to be operated in high ambient temperatures, install one of the aluminum pans as these do not hang much farther below the engine than the stock item, yet they add lots of cooling area. When you have installed the stock or special pan, trim off any gasket which protrudes at the front of the pan so that there will be no danger of the gasket preventing the transmission from bolting squarely to the adapter.

Install the new flywheel, using Loctite on the capscrew threads as the crank-flange holes open into the crankcase. Add the clutch, making sure that the depth from the mounting surface to the clutch surface of the flywheel is correct for the clutch which you are using. Porsche '63–'65 clutches can be used under certain circumstances, or use the 1500 VW clutch and disc for engines not exceeding 102 HP. 110 and 140 HP engines should be equipped with the Heavy Duty dune buggy clutches. Crown's competition pressure plate had nine extra springs and approximately 2,000 pounds. It is hard to push and more critical in adjustment because parts start to bend with this kind of pressure. Do not order it unless you need it (for engines producing in excess of 150 HP).

If you do not get your clutch disc and pressure plate from Crown, be sure to use the VW or Porsche parts which they recommend. Failure to do so will result in either a slipping clutch, or one which cannot be depressed because there are several critical dimensions which must be observed, including disc thickness, pressure-plate position vs. flywheel cavity depth, and throwout-bearing position. If you use the competition clutch, braze the pedal end of the clutch-cable housing tube to the chassis structure so that it will not be torn away from the chassis in operation. An eye bolt can be used to repair a loosened tube at the pedal end.

Adapting the Corvair starter merely consists of cutting off the shaft-bearing extension, retaining sufficient material for attachment of the Crown starter adapter. This adapter offsets the starter to permit mating the starter gear with the Corvair-pattern ring gear used on the Crown flywheel. Other adapter bolts have flywheel gear problems because the US-pattern starter gear does not mesh correctly with the VW or Porsche flywheel gear—and as a result, the flywheel is soon damaged beyond use. Because the starter is offset, an offset bushing must be installed into the VW or Porsche transmission case. This bushing is supplied with the starter adapter. Remove the stock bushing and install the offset bushing so that the starter shaft will be moved in toward the transmission mainshaft centerline. The bushing is notched and this notch is aligned with the rim of the case. Check the fit of the Corvair starter shaft in the new bushing. It may be necessary to ream the bushing to 15/32 inch to obtain the proper fit because the bushings sometimes close down when they are installed.

Single or dual-muffler installations can be used. The available area requires using small mufflers in the case of VW sedans and Porsches. All conversions work best with a rear motor mount. A rear mount for Porsches or VW sedans can be hung from a cross tube mounted between the bumper brackets. I used a VW side transmission mount for the rear motor mount, and suspended this from a piece of rectangular steel tubing spanning between the bumper brackets, For bus-type VWs, use the '59 and later Transporter rear bumper and the Crown Basic Rear Mount Kit. Trevor claims that this bumper is good, provided that you do not allow it to obstruct cooling air exhausting from the engine. He ducts through the bumper, or fabricates strip-steel bumpers to allow the heated air to escape. To permit bus and Transporter rear engine cover closing, use a saber saw to cut along the sides and bottom of the license plate, then bend the plate and sawed area of cover out 1-½ inches at the bottom to clear the back of the engine. Sedan deck lids also fail to clear the larger engine, and it is necessary to install Crown's bustle-bomb fiberglass rear deck lid extension, which could be ordered pre-gelcoated in standard VW colors. This cover is recommended both for neatness and to avoid recirculation of heated air through the engine. One big advantage of the Karmann Ghia body style is that the Corvair can be installed with no visible change to the bodywork. A built 140 motor in one of these would make for a real sleeper.

Last, but not least, you should seal the engine compartment so that none of the heated air which has been exhausted by the engine will be reused. You want only cool, fresh air in the engine compartment, If the heated air is allowed to reenter the engine, it will be difficult to cool the engine adequately. Do not neglect this important part of the installation. Take a look at a '65–'67 Corvair to see how the seal can be installed, Trevor recommends the use of neoprene diving-suit sheet rubber because it is easy to work with. A bit of cement, a pair of scissors, and an hour or so of work will seal any compartment.

Note strap-type rear motor mount, elimination of entire cooling system. Special ignition is not needed on Corvair engines.

Going Ghia VW owned by John Martin of Huntington Beach, California. Four-carb Corvair engine surprises Detroit monsters at stoplights and on freeways. Car runs 90 mph in quarter mile, yet delivers 23 mpg in daily drive-to-work use. Engine-compartment seals should extend to inside of rear bumper.

V8 and Other Engine-Swap Corvairs

The drastic step of upping the Corvair's power train to a V8 or Turbo V6 is popular for a number of reasons. Because the stock Corvair body is retained, the car does not attract the unfavorable attention that the "law" is known to give to special-bodied automobiles or sports cars. It becomes a street "sleeper," one that can be used to surprise other cars whose drivers might not expect such powerful performance from an "economy" car. In general, cars from the era of the Corvair are no longer subject to smog equipment requirements. Finally, the late-model Corvair suspension performs well in a high performance street application even with 500 HP onboard.

Several V8 Corvair kits have been produced. Nearly all of the "kits" involve locating the new motor in the back seat of the late model '65–'69 Corvair, at the time inexpensive used cars. Based on only hearsay evidence, about 1,000 of each kit were sold. The two basic designs can be referenced as the Crown design and the Kelmark design. MID Engineering used the Kelmark style adaptation. Neither Crown nor Kelmark is still in business, but the Crown design is still supported with some parts available from Clark's Corvair Parts. Several of the required Chevy parts are still available and even some aftermarket support is still around. The most common source today is partially finished cars from the past. The car was started back in the '70s, completed enough to run, driven enough to find out the flaws in the adaptation, whether in the design or installation, then parked for "a while." Those cars come onto the market via eBay or Craigslist, and find new owners. The original design purpose of both major V8 'vair designers was an inexpensive conversion using mostly available (at the time) parts. Since then, thousands of additional hot rod and street rod accessory items have come onto the market, in many cases providing builders with lots of different ways to build the cars. No two V8 Corvairs are the same. They are broad canvases that owners can make their own.

Kelmark Style

The Kelmark design involved rotating the Corvair transaxle 180 degrees so the input shaft points forward in the car, out of the differential. The new engine attaches to the differential and the input power flows past to the transmission, located behind the differential, then back into the pinion gear. There are a few problems with this design. First, the pinion gear is now driving the back side of the teeth on the ring gear. Even if readjusted to gain the correct tooth contact patch, the gears are not operating within their design parameters. Second, the long input shaft acts in some ways as a torsion bar dampening inputs to the transmission, but it is limited in its torque-carrying ability and is prone to failure. Third, with the shifter sticking out the back of the transmission, behind the engine, shift linkage becomes problematic. The benefit of hanging the trans behind the differential was increased legroom for the driver and passenger. Richmond Gear manufactured a limited number of "reverse-cut" ring and pinion gear sets. Use of these gears would allow a mid-engine application with the motor being reverse rotated while the transaxle gears rotated in their intended direction. The ring gear issue goes away with these gears. The others remain.

Ready for installation, this Kelmark-style powerpack shows how close the engine is to the differential assembly. In the Crown-style install, the transmission is in between.

Crown Style

In 1967 Ted Trevor of Crown Manufacturing began to develop and test a Sports GT car for the road: a mid-engine, full independent suspension, stock-appearing machine that would perform like a FIA Can-Am car made street-legal. Ted decided to use the 1966 and later Corvair because it allowed room for the mid-engine location, has a transaxle which not only turns in the correct rotation to permit coupling a V8 engine to the front but also shares transmission and other parts with the GM V8s to make a "bullet proof" conversion. The result is the Corv-8, not a vehicle for the novice driver, but a superbly balanced unit which, at the time, stood head and shoulders above any car you could buy that was supposed to handle and perform. After one ride in the prototype Bill Fisher decided to build one of these machines because it truly was the ultimate vehicle for the automotive enthusiast who wanted the very best—and at a reasonable price. You could build one of these "street sleepers" from your 1965–69 coupe, convertible or sedan for less than $1,000 with about 40 hours of work. It would cost more today, but many cars and kit parts appear regularly on the used market.

What is the actual performance potential? It's best described as "unreal." You have to "nail your goosebumps down with a hammer" after the first ride. With a 327/350 HP (stock cast-iron exhaust manifolds and quiet mufflers) the Corv-8 ran 105 mph in 12.22 seconds for a quarter-mile drag course. A high performance engine such as the Z28 302, LT-1 350, 396 or 427 should provide quarter-mile speeds of 115 to 130 mph.

Handling is superb primarily because Trevor invested more than $10,000 developing a new set of pickup points for the suspension and new front and rear sway bars. Spring rates were also revised with new springs at all four corners. The car goes around slalom or road-race courses "on rails." These same modifications work to keep the rear tires flat on the ground during hard acceleration by eliminating the swing-axle-like camber changing tendencies of the stock suspension. Weight, incidentally, is only 2,770 pounds with half a tank of gas and no spare tire—about 150 pounds more than a stock Corvair.

If you use a '65 Corvair then you will need a '66 or later transmission which is, internally, the same strong Saginaw gearbox that Chevy put behind many of their V8s. Only the complete transmis¬sion and a differential case need be pur¬chased because, with one exception, the differential internals are the same throughout the '65–'69 series. Consider a '65 for Corv-8-ing if you are buying the car specifically to build a Corv-8, as there are a lot more 1965 cars around. Besides changing from 4 to 6-bolts to attach the transmission and the differential housing, Chevy changed one other main part from 1965 to 1966. The pinion shaft of the '66 and later is slightly shorter. Trying to install an unmodified long '65 shaft into a '66 or later differential will result in an interference fit with the '66-up transmission, and eventual failure. See the drawing below.

The differential is the weakest point in the conversion. Unless you have a storehouse of late differentials, proper gear setup and some discretion in driving will be necessary to keep the car operational. The Corvair differential housing is, essentially, a box with big holes in 4 of the 6 sides, not a recipe for strength. A popular modification is to add a special top cover with match-drilled shoulder bolts holding it down. This will stiffen up the "box" and still allow disassembly. Cooling fins can also be added to it. During the assembly of the powertrain, it is important that the input into the bellhousing be carefully aligned. The center of the bellhousing can be dial indicated to the crankshaft to make sure of the alignment.

In the last few years two new conversion twists have been added. The 2-speed Powerglide transmission for the full size cars has been adapted to live between the V8 motor and the Corvair differential. The conversion involves removal of some of the auto-shifting capabilities of the PG, but results in a smooth operating, easy on the differential, setup. It also provides a Park position, never before seen on the Corvair! The kit was available from Clark's. Although ending up a bit longer

Crown engine room is covered by fiberglass seat form, shown here prior to installation of pre-sewn upholstery kit. Underside of fiberglass is coated with one inch of flame-proof polyurethane insulation to reduce noise level and heat in the passenger compartment.

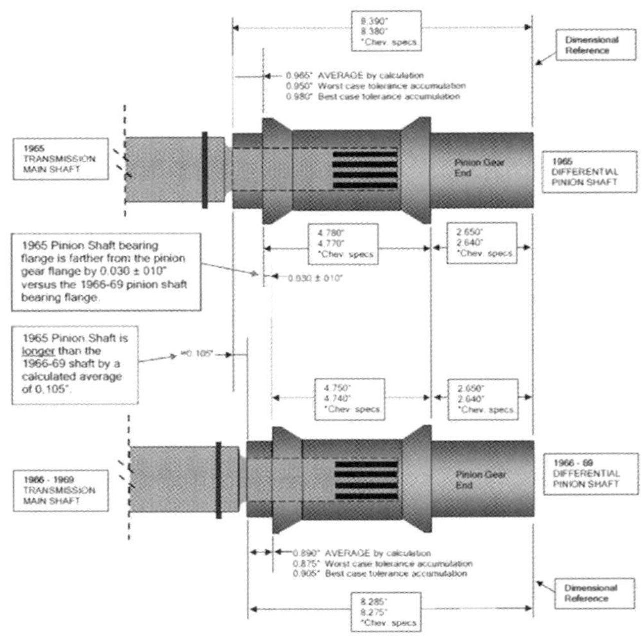

Info and drawing courtesy of Clark Hartzel, Bob Nichols, and Dan Drommerhausen.

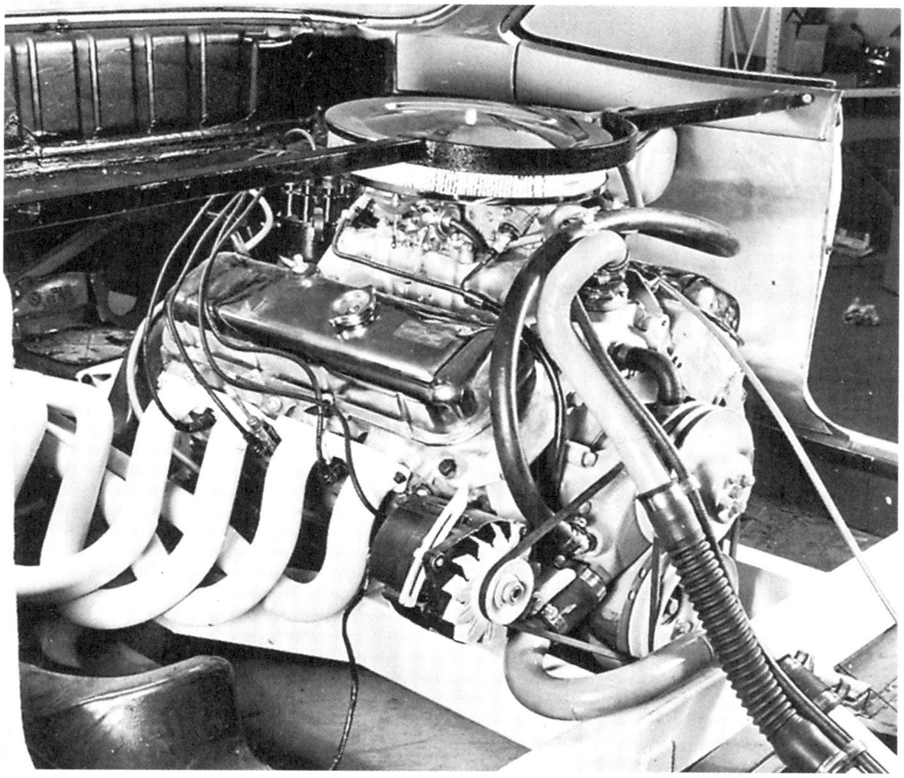

Strange home for a big block: the back seat of a Corvair '65–'69 coupe. Although the transaxle likes the torque of small block much better, big blocks have been used in the very well engineered Corv-8 conversions engineered by Crown Manufacturing's Ted Trevor. Even with a small block, the performance is very Group 7-ish!

215-cubic-inch GM aluminum V8 mounted in rear requires reverse rotating of the motor and some cooling system planning, but hardly increases the weight on the rear wheels and is a most impressive install!

than the original Crown installation, it is similar in layout. Another conversion using the regular Chevy Saginaw transmission case and a modified Hurst shifter has also recently hit the market.

Still in the prototype stages is the ultimate conversion kit for a mid-engined Corvair. It replaces the entire drivetrain and rear suspension of the car with a purpose-designed transaxle, based on heavy-duty racing parts and a choice of multi-speed automatic transmissions. The differential includes a quick-change-ratio system, so that a car can be set up for most applications, quickly and easily. This system is designed to handle more than 750 HP.

The cooling system has a hard job in any Corvair-to-water-cooled conversion. With the mid-engine location, there will always be a lack of natural airflow over the motor to externally cool it. Over the years, radiators have been mounted in every conceivable place. The most successful point is in the front of the car, with incoming "air management" and disposal of hot air during operation. But there are no sure things, many details have to be chased down and addressed. Other locations have been engineered and executed successfully. Kelmark originally recommended a rear radiator location.

The installation manual for the Crown conversion is still available from Clark's Corvair Parts. I recommend buying the manual to help you decide if the V8 path is the one for you. There is a specialty group dedicated to water-cooled Corvairs. It is known as the V8 Registry. The members are a specialty chapter of Corsa and put out periodic newsletters, and have both regional and annual national meetings at the Corsa convention.

A word of caution: any V8 Corvair project is a complex job. I have seen hundreds of conversions over the years. I've even owned a couple. The Crown conversion bears a close resemblance to a kit car, although it is much easier to complete than most kits. Many builders have a problem actually getting the car finished. They can get it 80% done, perhaps get it running, then drive it a bit, but never get all the bugs worked out. Those are often the cars that hit the market, having sat for years before the builder realized his limitations. As long as the purchaser realizes that the last 20% of the car will take longer than the first 80% did, success may be within his reach.

Other Conversions

There have been several other mid- and rear-engine conversions built over the years. Several Toronado powertrain swaps were built, there was even a kit to do it. This involved placing the complete longitudinal Oldsmobile power pack from the front of the Olds between the rear wheels on the late Corvair. Most of these conversions suffered from handling issues due to different length drive shafts side to side. The Corvair uses the driveshaft as the upper link in the rear suspension. The Oldsmobile was not designed to do this. Several Cadillac models were also available with this GM drivetrain layout and some Caddy-engined vehicles have been built.

The svelte body design and "right" size of the Corvair has also attracted some new swappers. Since the wheelbase of the Corvair is 108", the same as many '80s GM intermediates such as Malibu, Cutlass, and Monte Carlo, there have been several drop-on conversions built that, in essence, drop the Corvair body onto the frame/chassis of the water-cooled donor vehicle. Although these cars appear to be a Corvair, they are really RWD sedans. The good news is that all the engine and suspension hardware for the GM G-bodies will work, the drawback is that all the issues of the nose-heavy, solid-axle sedans carry over as well.

There have been many rear-engine swap vehicles built, too. There was a kit to install the 215 CID GM/Rover aluminum V8 in the rear. In addition home-built cars have been finished with both Buick and Chevy V6s. One of the hurdles to overcome with the rear engine kits is the engine rotation. All the other GM engines, indeed almost all other engines, turn the opposite direction of the Corvair, so replacements must be reversed to drive everything in the Corvair drivetrain. Several engines were also used in marine (boating) applications, where twin-engined craft required a pair of counter-rotating motors, so parts, like camshaft and seals, are available to "reverse-rotate" these motors. Some reengineering is needed of accessories, like water pumps and alternators, but it is doable. Many of those engines mentioned predate serious computer engine controls, but have aftermarket speed equipment available for them. If you decide to move away from the Corvair transaxle, perhaps to a Porsche unit, you no longer have the rotation issues, but other ones will pop up, such as finding suspension mounting points on the differential. The inventiveness of the Corvair hobbyist never ceases to amaze me, though, and several rear-engined, GM LS Series powered Corvairs are running around. I am sure they are amazing the people next to them at the stoplight as well. The rear-engine vehicles retain one of the Corvair's best attributes, a back seat!

Frank Parker's Northstar-engined Corvair shows what workmanship can do. The full Cadillac power train was installed and the body and frame adjusted to meet it.

Harlan Colburn of Denver, Colorado, built this mid-engined, Crown-style race car into a multi-time Cole Award winning autocrosser. The motor is a supercharged 383 Chevy V8, mounted a few inches behind the driver's right shoulder. The car is shown here winning fast-time-of-day at the 2010 Iowa Corsa Convention. This car also took fast-time-of-day at the 2011 Denver Convention.

During 2012, I inspected two of the most inventive engine swap Corvairs I have ever seen. Both replaced the Corvair suspension as well, so calling them Corvairs is a stretch. The first one was built by taking an air-cooled Porsche 911 4WD Carrera4 (that had been rolled), removing the body, fixing the minor mechanical damage, stretching the wheelbase, and stuffing it under a late model Monza body. The builder brought it to a Corvair club meeting and absorbed the attendees for about an hour. The Corvair guys were impressed with the quality of work needed to bring it to fruition. The builder received a little less enthusiastic reception at the local Porsche meet. The second inventive swap I saw was at the Corsa Convention in Massachusetts. On the day of the Car Display, this fellow drove in with his late model coupe. Aside from a slight extra flare on the front wheelwells, the car looked stock. As you got up close, you realized that the Corvair was dropped onto a 2002 Camaro chassis, with 300 HP and a 6-speed. The interior was blended to function like the Camaro, and it was hard to tell where one started and the other left off. Building of either of these cars was beyond the capabilities of most Corvair mechanics, but the average Crown build is well within our reach.

Interior Modifications

Much of the fun of driving a car comes from not only the performance of the car, but the way the car feeds back information into your body—the feel of performance. It can be hard to appreciate the handling of the car if you are sliding around on a slippery plastic bench seat, or spinning a bus-size steering wheel, etc. There are really four points where your body interacts with the car. You should seek to optimize those four points for your intended use. New cars have added all kinds of adjustability into the seats, the steering wheel, even the pedal assemblies to make the driver either more physically comfy—isolating him from the road, think Lincoln—or more tuned in to the car's performance—think Corvette. The points of contact are: 1) the seat, 2) the steering wheel, 3) the pedals, and 4) the shifter (assuming a manual transmission). In addition, visual inputs, such as the view out the windshield, mirror views to the side and rear, and view of the instruments are important for mental comfort. Make the car fit you, rather than the other way around. So we will address each point of contact.

Seats

Chevy decided that the Corvair would be a production test base for the "bucket seat in the sedan" movement. The Monza introduced many Americans to bucket seats. Those early buckets were more like split bench seats, but Chevy spent some real money in their construction. By the time of the late models, specifically '65 and '66, Chevy started cheapening out on the seats, and the '65–'66 seats offer much less support than the '64s. However, since it has been 40+ years since that era, many newer cars have been produced to serve as a supply line of better seats for our Corvairs. The Corvair seats had a great feature. They bolt down with four vertical bolts or studs. No fancy clips or slots—four points. That makes it quite easy to adapt in other seats. You will find, as you visit the auto recycling centers, that many other cars used different types of mounts. Tip—if you are looking for seats for your Corvair coupe or convertible, donor 4-door cars often have the seat backs fixed in place, so they can't flip forward for access into the rear area. Check before you pick a pair to install!

Aftermarket Seating

Lots of seats are now becoming available from the far east. It is certainly an inexpensive way to go. If you select fabric on your seats, instead of vinyl, it may be subject to more fading than the OEM seat fabrics. I have seen these fabrics fade out in a few months of sunlight. There are still several suppliers of seats in America. SCAT Procar sells a full line of seats. They even have the adapter brackets listed for the Corvair. The old line companies such as Recaro and Koenig are still around. But the junkyard has a great selection of seats, some with the adjustable features that will greatly improve your driving experience. As far as adapting in these seats, I have built flat-steel "H" style brackets that bolt down to the floor and up to the seat, pretty simple, safe, and only $15 in materials. Just lay it out on paper ahead of time and drill to fit.

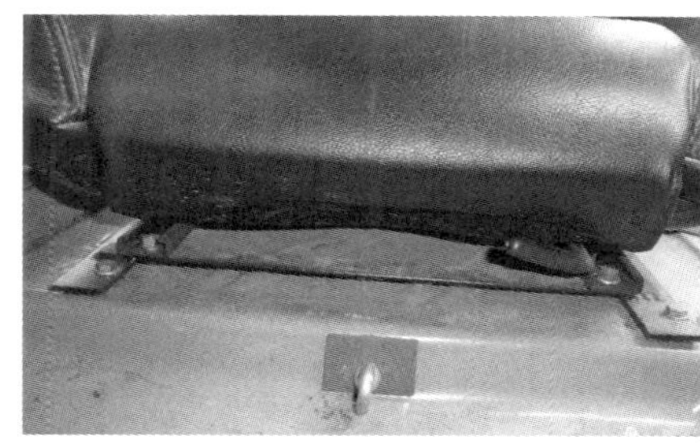

Most new seats come with sliding brackets with defined mounting holes. These flat adapters are strong enough for competition but thin enough to maintain headroom.

After removal of the original seats, measure the 4-hole pattern on the floorboard and lay it out on paper, then do the same to the underside of the seat bracket. Lay one bolt pattern on top of the other.

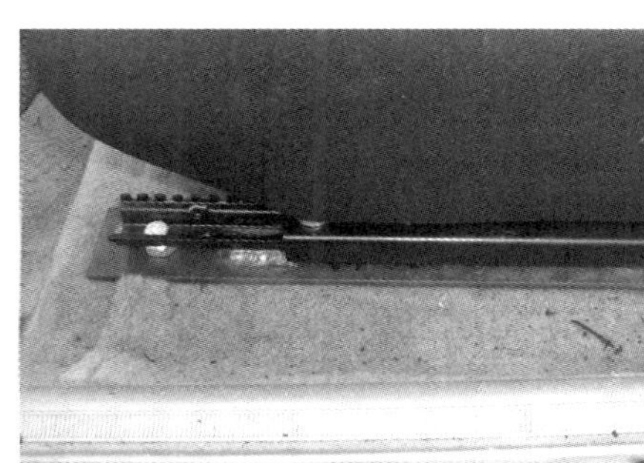

The only welding is on the cross connectors of the "H" brackets. The bracket bolts to the seat, then to the floor. Spacers may be needed to maintain clearance for the adjusters. Painting is recommended.

Note that the inboard rear mounting surface is lower than the other three. Be prepared. These mounts can also act to relocate the seat fore or aft to customize the range of adjustment for taller or shorter drivers.

Steering Wheel

I have been adapting other steering wheels onto my Corvairs since the '60s. At some point, I figured out that the Momo brand 6-bolt mount wheels used the same 6-bolt pattern as the GM wheels on the telescoping column and the "Sport Wheel" option. The bolt pattern is actually shifted 30 degrees around, with one bolt at the top, instead of a pair as on the GM design. If you want to just swap one onto a tele-column, you will have to make a couple of modifications to allow the locking mechanism to function correctly. But it is slick. I also learned that GM changed the steering wheel spline on the standard 1964 through 1966 Corvair to a smaller spline, in common with no other car. In 1967, to promote commonality with the other cars receiving the collapsible columns, they returned to the larger spline. Grant has made an adapter for the old-school 3-bolt style kustom wheels on the smaller spline. That adapter is now sold new only by Clark's Corvair. Grant still sells many wheels that work with this adapter. Since the 1980s Grant has promoted their "Signature Series" wheels. It is a larger 5-bolt pattern with a snap-in horn button design. All of their new designs use this pattern. The wheels come with a tall adapter sleeve and a rubber jacket, allowing you to bolt it onto a 3-bolt hub.

A cleaner solution is the direct-fit hub, now available for all years of Corvair. It uses the Grant 5-bolt pattern but is dual drilled with the GM 6-bolt pattern as well, widening your field of available steering wheels. As far as the wheel size and style, I caution you about going to a wheel much smaller than the stock (16-inch) diameter size. Most people can deal with a 15-inch wheel. I never recommend smaller than a 14-inch wheel, even for competition use. It depends, in part, on the remainder of the steering system: Quick box? Shorter arms? Fat tires? You may need the extra leverage of the larger diameter wheel. I prefer unpolished wheel spokes, that cuts down on bad reflections. I also like the feel of a thick leather wheel rim. Your desires may be different. That's why they make so many different wheels. The telescoping column was a popular option in 1965–66. The intended usage was to allow two different drivers of different stature to share a car, using the seat and wheel adjustability to match up. It has certainly worked that way on the racetrack where two different racers can easily share a car. Now coming out on the market are bolt-in, tilt-wheel steering column kits for the 1965–66 Corvair. These columns will also add 4-way flasher capability for cars that didn't have them as standard equipment.

The quick disconnect system for race cars recently became available for the small-spline '64–'66 Corvair steering column. The splined piece bolts on in place of the steering wheel, using the stock attaching nut. The second piece, the removable part, mounts to the steering wheel. A quick pull of the release ring and the wheel comes off for easier ingress and egress for the driver. This is especially handy when roll cage elements make climbing in a contortion. The kit is adaptable to most aftermarket wheels. For security, the wheel can be removed and stored in a tow vehicle.

For racing only—this quick disconnect bolts onto the original column.

With the new hub adapter, the Grant leather wheel looks at home in the Corvair.

Grant has dozens of styles available. This one is real mahogany

Any wheel can be adapted to the Grant pattern—Momo shown!

A great option for mature drivers dealing with roll cages.

This system is intended for racing use only. The Grant system for the 1967 Camaro will work on the '60–'63 Corvairs as well as the '67–'69 Corvairs.

There are now several new quick-disconnect steering wheel removal systems on the market. NRG, a Southern California company, makes units that fit in between an aftermarket 6-bolt wheel and an existing hub adapter. These units can be used as a security device on a regularly street driven car, because they do pass-through the horn signal. Some minor fit-up is needed to install these. There are less expensive units available on Ebay, but quality and fit may be an issue.

Pedals

Not much can be done on the location of the pedals in the Corvair. The only common modification I have seen involves extending the top of the accelerator pedal up and over to the left a little to provide an easier heel-toe relationship for the track. Make sure the rubber pads on the pedals are in good shape, replacements are available. If you don't trust the rubber, you can either glue on some heavy duty coarse sandpaper or paint the pedal with "traction" paint. It has coarse sand embedded in the paint. For competition use, you can add a dead pedal on the left front wheel well. It is a handy place to store your left foot at the track, between shifts.

Shifter

Most of the things that make the shifter feel right in a Corvair are not in the passenger compartment, so let's start with the transmission. If everything is okay inside, shifting is a matter of pushing, pulling, and rotating that shaft sticking out below the front crossmember. Doing that reliably, with little extra play and no hang ups is what makes the shifter a pleasure to use. Starting at the shaft, the stock coupling is made of three parts, plus the pin that aligns the coupler to the transmission shifting shaft. Two of these parts commonly wear, the pin and the shell of the coupler. The rubber inside the coupler gets hard, and sometimes will be attacked by oil and soften up, but that is rare. There are two holes in the shell that capture the pin to the trans. They are actually slots, a little longer than they are wide, allowing some misalignment during assembly. Even though the shell is hardened before assembly, the slots will wear over the years, adding slop to the shifter. An inexpensive way to renew the shell is to remove the coupler, drill out the slots to a small oversize round hole and add two shouldered bronze bushings between the holes in the shell and the pin. Because the shell is hardened, drilling it is tough. You can soften the material by heating it red hot and letting it air cool. This will really annoy the rubber in the middle of the coupler, to the point of smoke and flame maybe. Just be prepared. The only available bronze bushings for this application are too tall and must be shortened to fit. Once that is done, a replacement stock link pin and maybe a thin washer or two and the coupler is done. You can avoid this effort by using an aluminum replacement shift coupler from American Pi, a Corvair supplier. The coupler is retained inside the long shift tube from the shifter. A single clamp over the split tube retains the adjustment. If that clamp comes loose, you aren't shifting. Add a second stock clamp adjacent to the original and tighten them both.

On the late models, the shift tube that runs from the coupler up to the driver area is really three tubes. The inside tube does the work of shifting, the outer tube retains the hardware and acts as a fulcrum for the inside tube. That outer tube is retained to the transmission crossmember with a rubber-damped rod. Between the two metal tubes is a cardboard tube that acts as a stabilizing influence, keeping the inside tube from bouncing around (much). If water gets inside the shifter, that cardboard tube can get pretty ragged, causing the inside tube to hang up in fore, aft, and rotational movement. It can be very hard to remove and even harder to replace (in function). If you can find a shifter tube in good condition, start with that one. Chevy installed a plastic bushing in both ends of the outer shift tube, to retain the inner tube, allowing it to do its job. Those bushings always pop out, causing mushy shifting. If the car shifts sloppily, that is one of the problems. Clark's Corvair parts recognized this early on. Clark's first part number "C1" is a bronze bushing kit for the late shifter. Their kit is one way to address the issue. Don't just replace the stock bushings, they will soon fall out. Other bronze/copper/brass bushings can be pressed or soldered into place to fix that

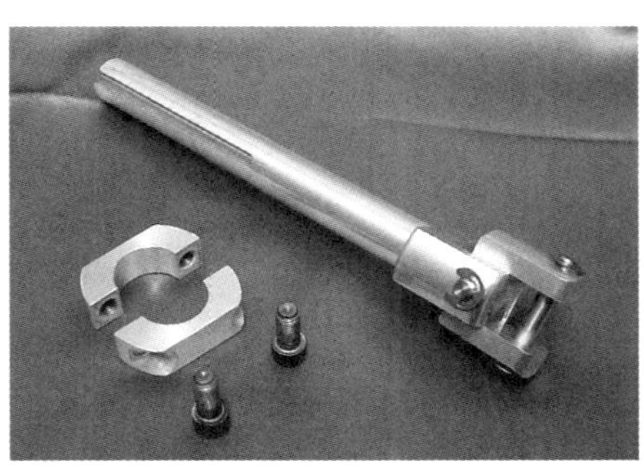

This billet aluminum shifter replaces the rubber coupling at the point of control of the shift rod into the transmission. It helps to greatly stiffen the response of the shifter.

Hurst shifter? With the Hurst lever grafted onto a '65 3-speed shifter, it shifts great!

Vinyl boot protects shift housing in dirty environment.

issue. Lube the inner tube well when you install it. Thick grease will make the tube sticky and slow the shifting, maybe use some 40 weight motor oil. The next part to check is the shim pack beneath the shifter. There are three items that need to be arranged to produce a tight, but smooth shift: the shift housing, the floor of the car, and the front stamped flange of the shift tube. The flange studs on the shifter have a shoulder where the shift tube flange bottoms out. The floor is clamped in between the shifter and the shift tube flange. GM supplied several different plastic and metal shims to remove the excess play at the floor. They can be mounted above and below the floor so that the shifter is solid to the shift tube and almost, but not quite, tight to the floor. The shift tube must be mounted solid to the trans, but not clamped tight to the floor. It should slide fore and aft with the transmission.

All Corvair floor shift levers are guided in their shift housing by two ¼-inch pins, one on each side. Years of driving will wear those pins flat on the sides, resulting in more lever play within the shifter. After you take the lever apart, it will be quite evident how to replace them. I have also seen guys just remove them, rotate them 90 degrees and reinstall them. Hey, it works!

Quick Shift Kits–Factory and Others

Factory "quick-shift" or short-throw conversion for '60–'64 Corvair 4-speed transmissions costs only a few dollars if your Monza or Spyder has bucket seats. Find part No. 3784295 (3-speed shift-lever housing), then clamp your 4-speed lever in a vise, using cloth to protect the chrome. Press the ball 29/32 toward the shift knob—keeping the slots aligned. Reassemble with the new housing and adjust linkage per section 6D in '62–'64 Shop Manuals. If you have a bench seat you will also need the 3-speed shift lever, Part No. 3812703. It installs without moving the ball. Shift lever travel is reduced 35% with these changes.

If you have a '65 or later 4-speed, the shift lever and housing from a 1965 3-speed (unique to that year) can act as a factory "quick shift." Unlike the early models, the fulcrum ball on the late-model shifter will not move easily on the shaft. It is not pressed into that location. (My guess is that it is brazed). The aftermarket quick-shift kits replace the stock shoulder bolts in the factory housing with longer, regular bolts. The modified housing

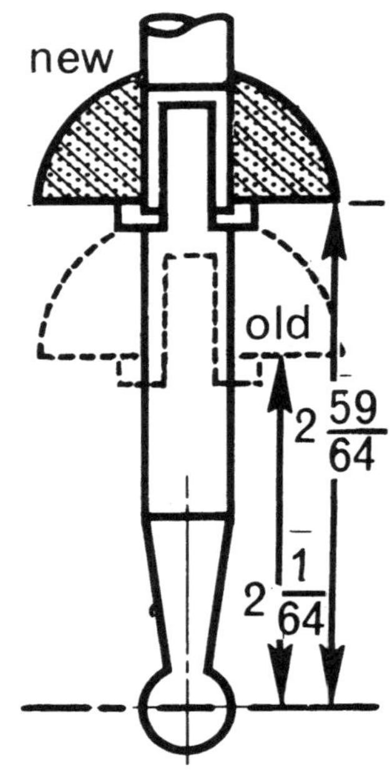

bolts down solid to the floor, and a small extension mounts onto the tip of the shift lever, locating the shift ball in just the right place. The change in the fulcrum point in the housing means the shift ball has a shorter distance to move, in order to push and pull the gear the same distance in the trans. The shorter distance however means the effort to move the shifter will increase.

22

Cars Gallery

Corvette Eater

Corvettes for lunch and XKEs for dinner was considered typical fare for this wild 1960 Corvair coupe. Concocted by Mike Jones of the Bill Thomas organization, this Anaheim, California-based machine ate up large machinery at the drag and road-race courses. For instance, it "shut down" 327-cubic-inch Corvettes at the 1964 National Drags in Phoenix, Arizona turning 101 MPH with an elapsed time of 13.5 seconds. On this particular occasion it obtained a new record "D" Modified Sports Cars and at the same time beat the "C" Modified Sports Car Corvette in runoffs. Since its construction in 1962, the car has given a excellent account of itself at the drag races, but often ran up against oil-temperature problems in the extended road-racing events. These were finally squelched by modifications to the oiling system as described below.

Numerous engines have been tried in the car, all of which were Corvairs in various stages of tuning. The 1964 engine consisted of Salih cylinders with Forgedtrue pistons using the '64 crankshaft and Spyder rods to obtain around 3 liters displacement (approximately 183 cubic inches). A hydraulic-tappet camshaft operated Corvette valves in the Thomas-modified heads with their usual four carburetors of the Rochester variety. The rectangular stack sticking through the blower shroud ahead of the fan uses a breather which opens into the engine cover through an 8-square-inch hole. Built-in tin baffles keep the oil droplets under control. Previously

tested breathers on the rocker covers and attached to the oil-filler tube were unsuccessful.

A heavy steel flywheel helps drag-race starts, with an aluminum Schiefer flywheel kept in readiness for road racing. A Chevrolet clutch couples the engine into the four-speed 3.89:1 ratio transaxle. Because this gear ratio is too high (too low, as a numerical value), 12-inch midget wheels are used at the rear for drag racing in order to get the ratio low enough to allow the engine to turn 6,000 RPM at 100 MPH as the car dashes across the finish. 5.00-12 midget slicks are used on the 7-inch rims. At the drags the fan belt is easily removed as it operates on one of the spring-loaded idlers. The belt is replaced for the trip back to the pit area.

Suspension is stock HDS at the front, with quick-steering arms and the Thomas sway bar. At the rear there is 2.5° negative camber with Thomas springs. Wheels are Hands and Halibrand. The driver compartment is starksville: all of the stock geegaws are stripped away to leave only two upholstered fiberglass buckets, a roll bar, Sun and Stewart-Warner instruments, Ray Brown belts and shoulder harnesses, and a Covico rubber-rimmed steering wheel.

Arizona Roadrunner

One of the fastest Corvairs ever was the '60 coupe built by Doug Roe of Mesa, Arizona. A list of the types and brands of cars it defeated overwhelmingly in hill climbs, slaloms, and oval-track racing is truly impressive. In addition to the Southwestern successes, the car and its driver were also entrants in the '64 New York 1,000-Mile Rally, and the Detroit Press-On-Regardless Rally. Wins include several fastest qualifier and overall firsts in sportscar oval-track racing at the Manzanita Park Raceway in Phoenix, overall first place at the Phoenix 1967 Annual Hill Climb, and first place at the Mule Mountain Hill Climb in 1966. Doug was regarded a maverick by Arizona sportscar types when he came to his job as Senior Development Engineer at the GM Desert Proving Ground. But his capabilities in engine building, tuning, and driving—plus an outgoing helpful attitude to all racers resulted in his election as president of the Arizona Four Cylinder Club. Doug also drove the '67 Chevy Corvair and Impala entries to impressive wins in the 1967 Pure Oil Trials. Photos here show the car with its owner, and include one pic of just some of the hardware won by the pair. Jerry Jensen, Doug's assistant mechanic and pit chief is the man on the right in the trophy photo. Louis Clements of NASCAR fame assisted Doug in setting up the chassis with a tunable front suspension that allowed jacking weight into either side of the car. The car's low stance was achieved by using springs with one coil removed: 283 Chevelle at the front, and HDS Corvair at the rear. Softer springs were used for dirt courses. Camber was negative all around: 1.5° in front, 2° rear. A four-speed transmission turned the 3.89 gear with Positraction and Greenbrier axles. Removal of all non-functional sheet metal and interior items reduced racing weight to 2,050 pounds.

The camshaft was a stock high-performance unit. Heads were 140 HP type with 10:1 compression. Special care was given to the assembly of the lower end, which used stock clearance for all bearings. Forgedtrue 3-9/16 bore pistons had drilled oil-return holes as Doug discovered that the usual slotted grooves are weak points in a piston. All of the oiling system was streamlined, with special attention to the entry and exit passages in the filler adapter, rear accessory drive, and block halves. An extra piece of metal welded to the pump pickup prevented losing oil out of the pickup in hard turns. The modified oil-cooler adapter had aluminum pipe fittings heliarced into it and the interior passages enlarged. This item connected to the Harrison stock car oil cooler. The crowfoot wrench was made from a 13/16 socket. It was used to torque the rocker stud/head nuts prior to every race. A stock oil cooler (12-plate) in the right-rear wheel housing cooled supercharger oil before it was returned to the engine. Gravity flow was used so that the turbo oil seal would not be overloaded. Doug's plans for the engine included using the supercharger to pressurize a center-mounted Holley four barrel feeding the heads by means of four runner tubes.

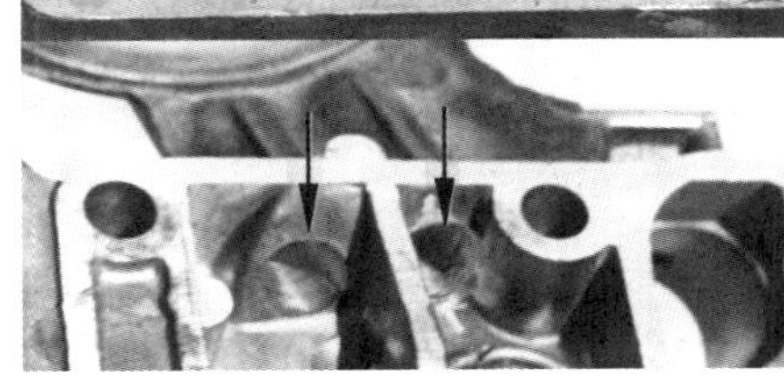

Third Life for an Old Race Car!

Bill Thomas and Mike Jones had built the original 1960 car under the direction of Chevrolet for small sedan racing in Southern California. It had originally been purchased for Bill's wife to drive. Because he had been a Chevy drag racer and Corvette tuner and already had a Corvair parts and modification business, it made sense for him to build it in his shop. Mike Jones drove for him. By 1964, Thomas was through with the car and Doug Roe, a GM engineer bought the car. He had just relocated to Arizona, and after his preparation of the Corvair for racing and hill climbs, went on to dominate many events in the region. This was the car's "second life." In the '80s, the car moved to California and Ed Connolly, with help from the Cactus Corvair folks, prepared the car for Laguna Seca vintage racing, but it was never quite completed for that life. When Warren LeVeque received the car he was tasked with preparing and restoring the car for vintage competition. Soon after, Warren was presented with the opportunity to purchase the car and the "third life" of the car really began. Warren has gone completely through the car, with an eye toward improving reliability and, at the same time, trying to maintain the appearance and flavor of the original design. The car is more than 50 years old! This car has competed at several events, and has been displayed at museums along with other Chevrolet performance milestones.

Doug Roe's turbocharged racer has been drastically modified with a front "snow-plow" which cleans up the car aerodynamically—and provides an unusual appearance.

Seth's Lola

The Lola T320 was originally built in 1972 with an air-cooled VW Type IV engine. It arrived at my home with a hole in the top of the cases with a rod from that VW sticking out. I chose this car for the Corvair engine conversion because the frame was already designed around a horizontally opposed motor. Most Formula cars are designed around inline motors and have frame rails where the cylinder heads of the Corvair would be. Because everything to the rear of the main roll bar bolts on, it was simple to build new rear subframes to cram in the longer Corvair motor. Because the Corvair motor was not to be reverse rotated, the Hewland was reversed to provide 4 forward speeds. It was simple until the mounting ears on the differential had to be swapped side-to-side as well. The car was originally built with a 45 DCOE Weber-carbed turbo motor. It was spectacular but difficult to drive. In turbo form, the car won the Don Yenko trophy at the 1990 Corsa Convention in Ontario. Shortly after, I converted the car to 40IDA 3-barrel Webers. Not quite as fast, but much simpler to drive. The car has 10" × 13" front wheels and 13" × 13" rears. The car won the overall title for two years in Northern California Council and SCCA regional competition. In 1993, the car, now with the Weber 3-barrels installed, again took the Don Yenko trophy at the San Jose Corsa Convention. In Ontario the car won the engineering class at the Concours and moved into the Senior Class. In San Jose, it repeated the Senior class title. In the mid-nineties, the car was sold to a racer in South Carolina.

Twice, this Corvair-powered Lola T320 took fast-time-of-day and won the senior division Concours class at Corsa Conventions. It now resides in South Carolina.

In early development the Lola Corvair had a modified 180-turbo motor installed. At just under 1,000 pounds—without driver—the car was a handful to drive.

Inboard Marine

Prop-walking is only one of the tests given Corvair-powered fiberglass runabouts manufactured by Inboard Marine of Temple City, California. Wayne Horning, well known for 12-port cylinder head designs for Chevy and GMC 6s, reengineered the Corvair for vertical operation. Exciting performance for cruising or skiing is provided by this 100 HP-plus family runabout. Smooth and quiet operation at all speeds is superior to outboards of comparable horsepower.

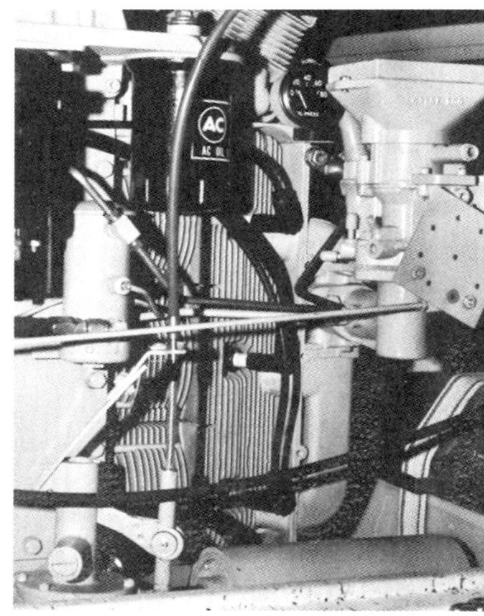

Bonneville

Water injection was also featured on the record-setting Spyder couple owned by Tom Keosababian of Whittier, California. The car turned 133 MPH at the 1964 Bonneville Nationals. Injection system was made by Gary Bailey. Externally mounted blower pressurized four carburetors on big-valve heads. Note careful attention to exhaust wrapping. Photo courtesy Spence Murray of *Popular Hotrodding Magazine.*

Sand Drag Corvairs

Photo by Lone Star Graphics.

Terry Glasscock of Belton, Missouri has a long history with his Top Eliminator sand dragster, Danger Zone. With 13 years of experience in this car, Terry is still winning at Thunder Valley sand drags in Missouri and around the Midwest. Terry and his father, Bill, have been drag racing since 1984, with seven sand drag season championships in just the past 15 years. The turbocharged, fuel-injected Corvair propels the sand car through the 100-yard course in as little as 2.86 seconds. The 3.1-liter motor runs on straight methanol and drives a 2-speed, air-shifted transmission.

Photo by Terry Glasscock.

Monza GT

Any discussion of Corvairs must eventually get around to the two fabulous GM "wish" cars that were produced as a result of the combined efforts of Chevrolet's Engineering and Styling groups. The coupe is the "wilder" of the two when you consider innovations. As such, it literally reeks with good ideas which you may want to consider for any special machine that you might be considering. A deep platform frame, similar to racing Porsches, using torsion bars with unequal length A-arms at all corners, rides on Halibrand magnesium wheels stopped by disc brakes. In the engine compartment ahead of the rear transaxle sits a supercharged engine. Apparently, various engines appeared in both of the cars, as engine sizes vary and so do the HP ratings which were quoted from time to time. Careful looking at the engines betrays numerous non-production components, so it is safe to assume that the cars were used for testing some of GM's new ideas. The engine gets its cooling and carburetor air from two ducts ahead of the rear wheels. When the rear portion of the body is hinged upwards you can notice that the hinge point centers around the twin-outlet mufflerless pipe which extends from both sides of the vehicle.

Torsion bars then become visible. These are bars with simple unsecured bent ends—nothing pretentious, but another good idea for special builders. The steering wheel is detachable and adjustable. Pedals are also adjustable for driver height. Headlights behind clamshell doors are one of the lesser ideas on the car as these are below legal height and would have been better placed in the fenders. A complete discussion of the GM show cars appeared in almost every magazine of stature, *Car Life* for May 1963, *Sports Car Graphic,* August 1963; *Road & Track,* August, 1963; and *Car Life,* December 1963. This car is still owned by GM styling and has appeared at several GM events in the last few years.

Astro 1 Concept Coupe

Chevrolet Division's ASTRO 1 coupe was introduced in 1967 as an experimental vehicle designed to combine the best styling efforts with automotive aerodynamic characteristics observed by Chevrolet engineers in wind tunnel testing. Beauty has been combined with a low-drag form. Frontal area—the major contributor to air resistance—has been reduced to a mere 13.9 square feet. Aerodynamic lift and yaw conditions which are often encountered, even at highway speed limits, are familiar to almost all drivers. A low penetrating nose counteracts lifting tendencies and the high tail with relatively vertical rear fender surfaces works toward eliminating wind wander in gusty conditions. Drag has been reduced by using flush releases and ornaments. The rear-engined layout also reduces the drag which is usually induced by internal airflow pressures. Only low brake-cooling inlets and cockpit-air intakes at the windshield base mar the smooth nose contours. The underbody uses unit construction with boxed sill members integral with the belly pan and bulkheads. Sheet metal of the floor and sill structure is liberally convoluted for stiffness. An anti-slosh rubber-cell gasoline tank in the right sill tends to offset driver weight. The bulkhead behind the seats extends above the seats to serve as roll-over protection in combination with the forged-aluminum windshield frame. The seat headrests include similar protection.

Double-wishbone front and rear suspensions attach to the underbody structure. Disc brakes at all four wheels are operated from a dual master cylinder. Brake cooling is provided by air inlets at the lower front of the nose and just ahead of the rear wheels. Panic braking actuates airflow spoilers at the rear, together with panic stop lights which are uncovered by operation of the spoilers. Two-piece cast-magnesium wheels have detachable outer halves to simplify tire mounting and to permit varying rim width easily. 5.50 rims at the front and 7.00 rims at the rear mount 5.50-13 and 7.00-13 Goodyear Sportscar Special Tires. Front and rear treads are 54.7 and 55.9. The car has an overall length of 176.7 inches, width of 72.2 and height of 35.5 with the canopy closed.

Perhaps the primary "gee-whiz" aspect of the Astro 1 is its seating arrangement. Push on either of the flush canopy-release pads and the canopy is raised electrically. Seats are thereby brought to a near-vertical position. You then step in over the sill onto the non-skid surface of the seat platform. Touch an interior handle and the assembly closes, moving you and your passenger to a semi-reclining position

where you can adjust foot pedals and twin-grip steering wheel to your liking. Energy-absorbing steering column and inertia-reel and quick-release harness systems are also included.

Instruments include a 160 MPH speedo, 9,000 RPM tachometer, oil pressure, alternator, cylinder-head temperature, canopy ajar, seat unlocked, and seat-belt unfastened.

Chevy enthsiasts wished that GM had introduced both the Monza GT and the Astro 1—which demonstrated what can be done when the imagination of engineers and stylists is allowed to roam free.

Genie-Corvair

The Genie-Corvair race car was originally constructed by Huffaker Race Cars in San Francisco, California, for Mr. John Grow of Friendly Chevrolet in Rialto, California. A parade of engine problems, mostly concerned with high oil temperatures, proved once again that it is often less expensive and far less frustrating to race with an engine specifically designed for that function. Converting an auto engine can be done, obviously, but the money merely goes out in smaller bundles—to end up at a similar sum. However, it should also be pointed out that when pioneer efforts are carried out—as in the case of this car and engine—men who follow in similar paths later on can often go farther on less money. This statement assumes that the developers share their hard-earned and expensively obtained knowledge. Mr. Grow has done just this, and most graciously.

High oil heat was the original engine killer. The stock oil cooler was completely inadequate. Another cooler was tapped in, but offered no improvement. Oil temperature of 265° F was looked on as "cool." Examination of the installation showed that the cooler, in the same line with the filter, created sufficient restriction so that the internal by-pass valve opened to short-circuit the oil around the cooler. Cooler routing was subsequently set up to eliminate restrictions and by-pass valves. The final installation was a dry-sump oiling system which used 2.5 gallons of oil in a dry-sump oiling system in conjunction with two Spalding-type oil coolers and an extra pump handling dry sump and cooler circuits with a 409 pump adapter taking care of the engine oil pressure. Keeping the belt in place at high RPM was also a problem. This was solved by Jay Houtz, who handled engine-construction chores on the car in addition to his used-car reconditioning activities at Friendly. Jay spring-loaded the idler and belt problems disappeared. There were also airflow-ducting problems which proved that it was a good plan to leave the factory ducting in place and provide sufficient air-tight inlets and outlets to ensure that hot-air recirculation cannot occur. Any builder planning a car with the engine ahead of the transaxle should remember that this can be a problem.

After numerous efforts, Houtz assembled two completely reliable engines for the car. Basically the two engines are almost identical, and both use parts from previous engines. 1964 lower ends are used, with the 2.94-inch-stroke cranks chromed, and the shot-peened connecting rods equipped with 327 bolts. Salih cylinders are used. One engine has Jahns pistons, the other has Salih pistons. Camshaft has timing similar to the '64-'65 high-performance models, but with added lift. Hydraulic tappets are used. Four-carburetor heads are the Thomas big-valve models which are similar to those used on the 140 HP Corsa. Both engines crank out 165 HP at the rear wheels, which is probably around 200 actual ponies. Grow and Houtz felt that this was the maximum which could be expected from the engine unless the cases were made more rigid so that the halves did not twist against each other at high RPM and high HP outputs. They discovered that power beyond this output makes engines "too exotic" and are not likely to last out a race. The engine exhausts into a handsome tuned system. At the front of the engine there is a Motorola alternator. At the rear (clutch end) a large breather is mounted to the engine cover or blower mounting. This latter item has also proved helpful in getting the oil temperature into the usable range.

In August of 1964 the car was purchased by Ed LaMantia, service manager for Rod Carveth in San Carlos, California. Ed ran the car at the Fall Laguna meeting to obtain another class first for the machine. The top left photo shows the car on that occasion. Note that the air intake is now of the Porsche type, which is perhaps done to match the Porsche 741 transaxle unit.

In 2010, the car was restored at the Huffaker Engineering shop, now located at Sonoma Raceway in California. It has since competed in several vintage races in a "revised" form. Now sporting Weber 3-barrels and a few other updates.

1964 Photos of the car on the track at Riverside are courtesy of *Sports Car Graphic's* Bob D'Olivo.

Updated version, 2013.

Rick Norris's Corv-8

Rick Norris writes about his V8 Corvair:

Back in the seventies when I first heard about the Crown mid-engine V8 conversion I wanted to build one. It was many years later before I got the chance. I found a complete Crown conversion done back in the day on a 1966 Monza coupe. It was for sale and I bought it. It had however been sitting unused in an unheated dirt floor garage in Chicago and was otherwise a solid project car with which to begin my dream of building one. Over the next ten years I rebuilt, replaced, and changed everything to my vision of what it should be. Needless to say it went far beyond where I originally thought it would. I used a 350 crate motor as a base, installed Edelbrock aluminum heads, roller rockers, and aftermarket port fuel injection. The dynamometer showed 340 HP. Cooling is a front mounted Griffin aluminum crossflow radiator. Final drive is 3:55.1 Positraction with '66 Corvair ultra close ratio four-speed transmission. Rear suspension is a four link, Crown anti-roll bar, and Gabriel adjustable shocks. Front suspension is modified Corvair with adjustable shock coil-overs with anti-roll bar and a Flaming River quick ratio steering box.

Rick sold the car to a California owner in 2007. A few years later, the car returned to the East Coast.

On a visit to Daytona Beach, Florida, Rick stopped by Smokey Yunick's garage.

The 350 found a comfortable home in the backseat of the Corvair.

Other Great Cars

Because it was just the right engine size, the Corvair was a popular midget racing motor in the early 1960s. Midgets were about the only front-engined, rear-wheel-drive Corvair-powered vehicles.

This Corvair powered sports racer, with owners Dan and Sandy Cole driving, won several National SCCA Solo II Championships. It has a B&M blower mounted on top of a modified 140 and uses a Hewland FT200 transaxle. It now resides in Virginia.

Note the license plate on this Ohio Corvair autccrosser.

Amazing amount of work involved with this changeover. That's a complete Porsche *Carrera 4* (4WD) chassis underneath the Corvair coupe body. The interior is a combination of the two but the car is all Porsche underneath. Beautiful workmanship.

At the Corsa drag races, part of the 2004 Lexington Convention, Billy Bruce's 1960 coupe squares off against Paul Fox's late model. Billy dropped into the 12-second range and had the fastest time of the day. The North East Corvair Council now hosts drag races as part of the Corvair Olympics every few years.

Billy Bruce of Tyrone, GA runs this early model drag car. It has run in the 12-second range. For traction and rear weight distribution the battery is at the very rear of the car.

Ray Clayton has pulled 380 HP out of this normally aspirated 3.1L Corvair motor.

Resources

Back when the Corvair was in production, many mainstream hotrod and performance suppliers had a section of their catalog devoted to Corvair items. Being in the performance business, not the Corvair business, many of them have gone on to support mostly the current automotive makes and models. Some of those manufacturers may still make a few parts that will fit our Corvairs, but hardly advertise the parts or services.

Fortunately, since the halt of the Corvair's production, several companies from across the nation have picked up the mantle of parts support. These vendors have looked at the Corvair parts market as a viable place to invest their time and money to obtain, reproduce, and supply the parts we need to keep our Corvairs on the road. In addition, many people around the country have specialized in a particular facet of the Corvair performance market, in many cases adapting parts originally intended for other cars to work with our Corvairs. Even beyond that there still exists a large group of mechanics that specialize in the Corvair and its repair and service. If you are lucky enough to live near one of them, congratulations!

Please note! This is only a partial list. The Corvair Society of America maintains a list of vendors for their members' use. Vendors and shops must be evaluated and approved before being listed by the club. They maintain the list and supply it to their members in their monthly magazine, *Corsa Communiqué.*

American Pi
19111 Chase Street
Northridge, CA 91324
(818) 349-9508
www.american-pi.com

Bill Cotrofeld Automotive
3630 Route 7A
PO Box 235
East Arlington, VT 05252
(802) 375-6782
www.cotrofeldautomotive.com

Black Hawk Engineering
737 W. 10th Street
Loveland, CO 80537-4624
(970) 461-1328
www.blackhawkengr.com

California Corvairs
15090 La Palma Drive
Chino, CA 91710
(951) 284-3555
www.californiacorvairparts.com

Clark's Corvair Parts
400 Mohawk Trail
Shelburne Falls, MA 01370
(413) 625-9776
www.corvair.com

Corvair Ranch
1079 Bon-ox Road
Gettysburg, PA 17325-8312
(717) 624-2805
www.corvairranch.com

Corvair Repair LLC
2214 293-1/2 Avenue NW
Isanti, MN 55040
(763) 444-9334
www.corvairrepair.com

Corvair Underground
PO Box 339
Dundee, OR 97115
(503) 434-1648
www.corvairunderground.com

Handy Car Care
1896 S. Gregory Road
Fowlerville, MI 48836
(248) 613-8586
www.corvairmechanic.com

John's Corvair Parts
1264 Walnut Street
Stoneboro, PA 16153-2516
(814) 336-9033
www.johnscorvairparts.vpweb.com

LeVair Performance Products
4627 N St. Rd. 9
Anderson, IN 46012
(765) 643-9290

Mike's Corvair Parts
Mission Viejo, CA
(949) 294-5699
www.mikescorvairpoarts.com

Paul's Corvair
805 N. Gray Street
Indianapolis, IN 46201
(317) 636-0154
paulvair@yahoo.com

Performance Corvair,
Silicone Wire Systems
3462 Kirkwood Dr.
San Jose, CA 95117-1549
(408) 247-2237
www.perfvair.com, www.wireset.com

Rafee Corvair
2695 Bengal Road
Wister, OK 74966
(918) 753-2486
www.rafeecorvair.com

Rear Engine Specialists
16010 W 5th Avenue Unit 12
Golden, CO 80401
(303) 278-4889
www.rearenginespecialists.objectis.net

SC Performance
Jeff Ballard
(805) 320-9032
JBYenko77@aol.com

Stinger Motorsports
2736 Calloway Street
Roanoke, VA 24012
(540) 204-0917
Yenko66@cox.net

Vairfast Fabrication
11604 New Bond St.
Fredericksburg, VA 22408-1864
(540) 891-9466

Wolf Enterprises
32605 Rhode Island Avenue
San Antonio, FL 33576
(352) 588-0645
gyoungwolf@earthlink.net

Recommended Reference Books

Herb Adams
Chassis Engineering: Chassis Design, Building & Tuning for High Performance Handling
ISBN: 1-55788-055-7, HP Books

CORSA, Corvair Society of America
Corvair Basics: an Introduction to the Unique Technical and Mechanical Aspects of 1960–1969 Chevrolet Corvair
Available on CD-ROM

Corvair Technical Guide Volumes 1 & 2

Corvair Technical Guide Volume 3

Richard Finch
How to Keep Your Corvair Alive

How to Air Condition Any Corvair Vehicle

Converting Auto Engines for Experimental Aircraft

Tony Fiore, CORSA
The Corvair Decade: An Illustrated History of the Rear Engined Automobile

Bob Helt
3016 Pearl Harbor Drive
Las Vegas, NV 89117
bobhelt@aol.com

Classic Corvair
A comprehensive technical description of the Corvair and its features. Contains maintenance and upgrade details. 330 pages.

Corvair Secrets
Things you probably don't know about the design and operation of the Corvair automobile. 250 pages.

How to Identify and Rebuild Corvair Rochester H and HV Carburetors
Comprehensive non-turbo carburetor rebuild information. 110 pages.

How to Identify and Rebuild Corvair Carter YH Carburetors Used on Corvair Turbocharged Engines
Comprehensive turbo carburetor rebuild information. 90 pages.

Mike Knepper
The Corvair Affair
ISBN: 0-87938-066-7, Motorbooks

Warren Leveque
Solo Is … (Collection)
Available from: Corvair Underground
Or the author:
Warren LeVeque, 4657 N. State Road 9, Anderson, IN 46012

Hugh MacInnes
Turbochargers
HP Books

Julius Mackerle
Air-Cooled Automotive Engines
ISBN 0-470-56290-0
1972 Revised Edition, John Wiley & Sons, Inc., New York

Dave Newell
The Incompleat Corvair
1979 (no longer in print)

Fred Puhn
How to Make Your Car Handle
HP Books

Ocee Rich
Corvair Guide
Vintage book (no longer in print)

Jim Schardt
Yenko Stinger Preparation Guide
(no longer in print)

Carroll Smith
www.carrollsmith.com/books
Carroll Smith's Nuts, Bolts, Fasteners and Plumbing Handbook

Prepare to Win: The Nuts and Bolts Guide to Professional Race Car Preparation

Tune to Win: The Art and Science of Race Car Development and Tuning

Engineer to Win: The Essential Guide to Racing Car Materials Technology

Drive to Win: The Essential Guide to Race Driving

All books are highly recommended.

Recommended Periodicals

CORSA Communique
by Corvair Society of America

Grass Roots Motorsports
by Grass Roots Motorsports

Hot Rod Magazine
by The Enthusiast Network

Super Chevy Magazine
by The Enthusiast Network